# CONSTRUCTING AUSTRALIA

ABC
Books

FILM AUSTRALIA

A COMPANION TO THE ABC TV SERIES

# CONSTRUCTING AUSTRALIA

### RICHARD EVANS AND ALEX WEST

THE
MIEGUNYAH
PRESS

**The Miegunyah Press**

The general series of the
Miegunyah Volumes
was made possible by the
Miegunyah Fund
established by bequests
under the wills of
Sir Russell and Lady Grimwade.

'Miegunyah' was the home of
Mab and Russell Grimwade
from 1911 to 1955.

THE MIEGUNYAH PRESS
An imprint of Melbourne University Publishing Limited
187 Grattan Street, Carlton, Victoria 3053, Australia
mup-info@unimelb.edu.au
www.mup.com.au

First published 2007
Text © Richard Evans and Film Australia 2007
Design and typography © Melbourne University Publishing Ltd 2007

**Warning** Aboriginal and Torres Strait Islander people should exercise caution when
viewing this publication as it may contain images of deceased persons.

Cataloguing-in-Publication data is available from the
National Library of Australia

ISBN 0-522-85402-8
ISBN 978-0-522-85402-2

Designed by Nada Backovic Designs
Typeset in Avenir by Nada Backovic Designs
Digital reproduction by Splitting Image
Printed in Australia by Australian Book Connection

# CONTENTS

Preface     vi

The Bridge     2

Pipe Dreams     66

A Wire through the Heart     120

Still Building     152

Visualising History     154

Recommended Reading     160

Acknowledgements     162

# PREFACE

Just as a river is made up of individual droplets of water, the past brims with many individual lives. When the actions, experiences and influences of all those lives are put together, the bigger picture we call history emerges. In Australia all those lives, and the history they created, have endowed us with a unique and distinctive culture and country. Constructing Australia is the story of three important turning points in that fascinating process. It tells the story of the building of the Sydney Harbour Bridge, the laying of the Overland Telegraph Wire and the construction of the Goldfields Pipeline running from Perth to Kalgoorlie.

Though very different structures, each is an engineering feat that fundamentally changed Australia's identity and its relationship with the rest of the world. Each structure is uniquely Australian; each was developed to solve uniquely Australian challenges of distance, geography, environment, politics and society.

The telegraph, the pipeline and the bridge were all unprecedented technological marvels on an international scale. In their day, no telegraph crossed such hostile country, no pipeline was longer or more advanced, and no bridge larger. They were all built well before Australia had even produced its first car.

People are always at the heart of great events and *Constructing Australia* interrelates these big-picture elements with the human stories at the centre of them. They are stories of struggle and suffering, of visionary plans and ideological conflicts, of personal successes and human tragedy. They show how very different and often conflicting characters can be thrown together and somehow achieve greatness.

The Overland Telegraph was built in the 1860s and 1870s by Charles Todd—a disciplined, religious, technology-obsessed engineer. He would never have realised his dream of connecting Australia without John McDouall Stuart—an alcoholic loner whose genius was in bush survival. Stuart became the first European to cross the torturous continent and live to tell the tale. The telegraph was the first step in ending Australia's isolation from the rest of the world—and within its own borders—but had a profound and devastating impact on indigenous people whose territories were soon overrun by white incursion.

Sixty years later the Sydney Harbour Bridge became a reality because another tenacious yet conservative engineer—J. J. C. Bradfield—found that his social vision for Sydney matched the political ambition of fiery socialist Jack Lang, a man who still excites passions to this day.

In between these landmark moments, another unlikely partnership in Western Australia—between a technocrat, Charles O'Connor, and a politician, Sir John Forrest—led to the creation of the Goldfields Pipeline. It would eventually cost O'Connor his life.

Each of these projects was revolutionary in its own right. Without the telegraphic connection to the world, Australia would have been stifled by intense isolation. It allowed the country to join the world and be influenced by events beyond the seas. In turn, government from London could be more direct. In 1900, in the wake of the construction of the pipeline, British colonial secretary Joseph Chamberlain urged Western Australia to join the Federation 'at once', thereby uniting all the rival colonies as a single country. The order came by telegraph and quickly assured that Federation would occur with Western Australia as part of the new nation.

It was a nation that a little over a decade later would be called upon (again by telegraph) to send its young men across the planet to fight in a world war. In 1860 it had taken Australia months to learn of the outbreak of the American Civil War. In August 1914, the country knew it would join the fight against the Germans within hours of the conflict erupting.

After World War I and on through the depression in the late 1920s and 1930s, Australia sought to come to terms with the terrible human, social and economic cost of war. The impact of this would influence the subsequent decision to move ahead with the construction of the Sydney Harbour Bridge as a symbol of hope. That Australia was so indebted to Britain after the war played a key role in how the political events surrounding its construction unfolded. It was a time of social upheaval and economic hardship, out of which the Sydney Harbour Bridge emerged as a potent national and international icon.

History is full of interesting connections of this kind, and *Constructing Australia* seeks to show how Australia's development was never just a straightforward story of how telegraph poles are erected or steel bars riveted together.

It brings us back to the idea of history being like droplets of water in a river. It is made up of individual events that are interconnected to create a wonderfully rich big picture.

**Alex West**
Executive Producer
Constructing Australia

# The Bridge

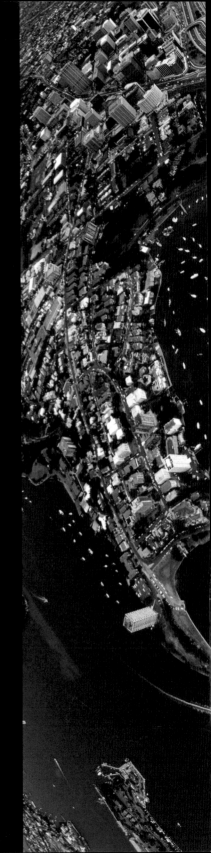

For 75 years it has been the symbol of a city, the vast man-made icon of the nation. Long considered a marvel of construction, the triumph of the Harbour Bridge is not its engineering—the real miracle is that it was ever built at all.

Lost in the decades since it opened is the true story of the bridge—how once it was the emblem of a divided country, driven to the brink of unprecedented violence.

Back then the bridge was the battleground for Australia's future . . . today it's a reminder of how close we came to a national tragedy.

# 1932

There was this disastrous situation with the imminent breakdown of law and order . . . there was going to be economic collapse.

ANDREW MOORE

# The players

## J. J. C. Bradfield

John Job Crew Bradfield was born in Queensland on Boxing Day 1867, the ninth of ten children. His parents were working class, and had migrated from England to Australia ten years earlier. Early in his education, Bradfield showed exceptional ability in mathematics. He won scholarships and prizes that allowed him to study engineering at the University of Sydney.

In 1891 he joined the New South Wales Department of Public Works as a draftsman. After working on several important projects, including the Burrinjuck Dam, he was promoted to assistant engineer. At public hearings in 1912, he proposed a suspension bridge across Sydney Harbour, linking Dawes Point with Milsons Point. It became a monumental engineering feat to which he would devote extraordinary effort and energy for the next twenty years.

Short and quiet by nature, Bradfield was unaffected in his personal relationships. However, he had a strong belief in his own ability and a talent for publicity that alienated some. He was hailed as 'the world's greatest engineer' in 1924; it was not surprising then that critics called him arrogant. More than once he was accused of claiming credit for the ideas of others. There was some truth to this, but Bradfield's genius lay in taking the half-formed ideas of others and incorporating them into a broader plan, and in inspiring others to help make the plan a reality.

Bradfield's second name was appropriate. Like the biblical Job, Bradfield showed an unshakeable faith in his ideals, despite daunting obstacles and seemingly endless setbacks. Although his vision sometimes over-reached itself, straying into the grandiose, Bradfield's commitment to a modern and unified transport plan for Sydney marked him as one of the great engineers of his age.

# Jack Lang

One of the most controversial and colourful figures in the history of Australian politics, John Thomas (Jack) Lang dominated the Australian Labor Party in New South Wales for more than a decade. Born in 1876, Lang experienced poverty and deprivation as a child when his watchmaker father became too ill to support his family. Lang became a real estate agent and auctioneer in the new Sydney suburb of Auburn, and prospered as the area's population grew.

Lang was physically imposing, 6 feet 4 inches (193 centimetres) tall and solidly built, with a big moustache and jutting jaw. He was not an eloquent speaker, but his harsh style was effective. Involvement in local politics took him into state parliament, and he won the seat of Granville for Labor in 1913.

When Labor took government in 1920, Lang was appointed Treasurer. During a period in opposition, he successfully campaigned for the Party leadership, which he won in 1923. Two years later he led Labor to a narrow election victory, and became Premier. Lang's government is chiefly remembered for introducing child endowment payments, widows' pensions and a workers' compensation act. His ability to make significant change, however, was restricted by a hostile Upper House. Labor lost the 1927 election, but Lang remained as leader, dominating the party to such an extent that his enemies called him a dictator.

In 1930, as the Depression began to bite, Lang swept back to power with a huge majority. However, as the economic crisis became more severe, Lang's policies became increasingly erratic and provocative. He split the ALP in New South Wales, and brought down the federal Labor government of James Scullin in 1931. With the economy and administration of the state in disarray, he was dismissed by the Governor, Sir Philip Game, in May 1932.

At the subsequent election, Labor was beaten by Bertram Stevens's newly formed United Australia Party. Lang remained Labor leader until 1938, but was never again a major force in politics.

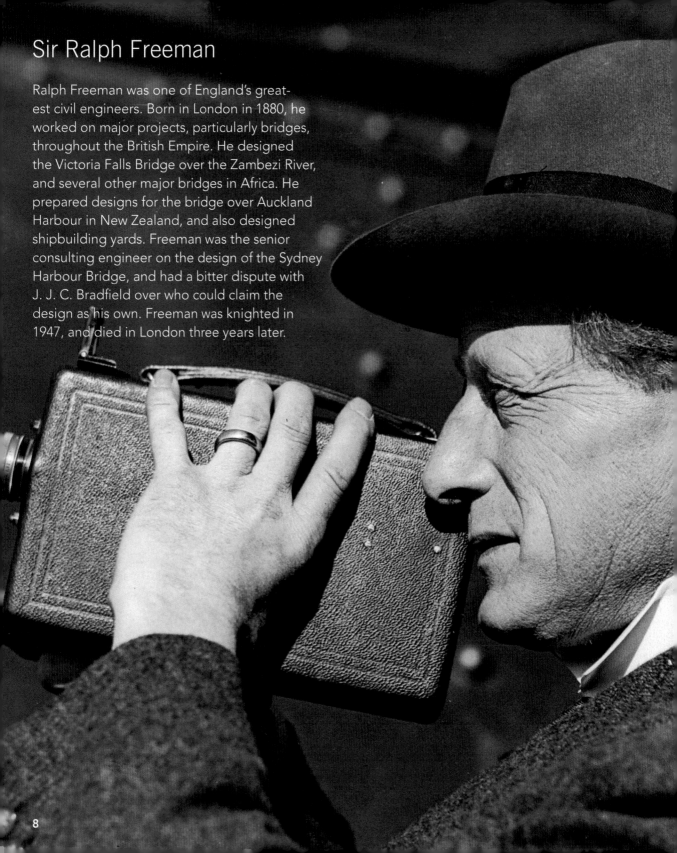

# Sir Ralph Freeman

Ralph Freeman was one of England's greatest civil engineers. Born in London in 1880, he worked on major projects, particularly bridges, throughout the British Empire. He designed the Victoria Falls Bridge over the Zambezi River, and several other major bridges in Africa. He prepared designs for the bridge over Auckland Harbour in New Zealand, and also designed shipbuilding yards. Freeman was the senior consulting engineer on the design of the Sydney Harbour Bridge, and had a bitter dispute with J. J. C. Bradfield over who could claim the design as his own. Freeman was knighted in 1947, and died in London three years later.

# Francis de Groot

Francis de Groot was born in Dublin in 1888. After serving with distinction on the Western Front during World War I, he moved with his wife to Sydney in 1920, where he established a successful antiques business. Politically conservative, de Groot was alarmed by the election of Jack Lang as Premier in 1930, and joined the New Guard, a semi-fascist paramilitary movement led by Eric Campbell. De Groot became famous for subverting the official opening of the Sydney Harbour Bridge in 1932: mounted on a horse and dressed in a cavalry uniform, he was able to cut the ceremonial ribbon with his sabre before Premier Lang. After serving in the armed forces during World War II, de Groot returned to Ireland, where he died in 1969.

# An accidental city

If someone were asked to choose a site to become Australia's largest city, it would not be Sydney. The soil is poor, the Blue Mountains are fire-prone, and the water supply will always be a problem.

Most of all, there is just not enough room: Where on earth will we put the airport? The available land is a tangle of long, awkwardly shaped peninsulas: transport infrastructure is going to be an expensive nightmare. Sydney Harbour, true, is strikingly beautiful: all the more reason to make it a national park, and build our city somewhere else.

But Sydney became Australia's first city because it had an excellent harbour. By the time the other disadvantages were apparent, it was too much trouble to move. Sydney grew, haphazardly, from a threadbare penal settlement to an important trading port and a commercial and administrative centre.

By the second half of the nineteenth century, Sydney's most pressing problem was the very reason for its existence: the deep-water harbour, which cuts the city in two. A vast fleet of ferries ran day and night, carrying passengers, vehicles and livestock from shore to shore. The busiest routes were those making the short crossing from Circular Quay and Bennelong Point to the North Shore. The volume of traffic was enormous. In 1890, the ferries carried 5 million passengers; by 1910 the number had grown to 13 million. At peak times, 75 ferries would dock at Circular Quay each hour.

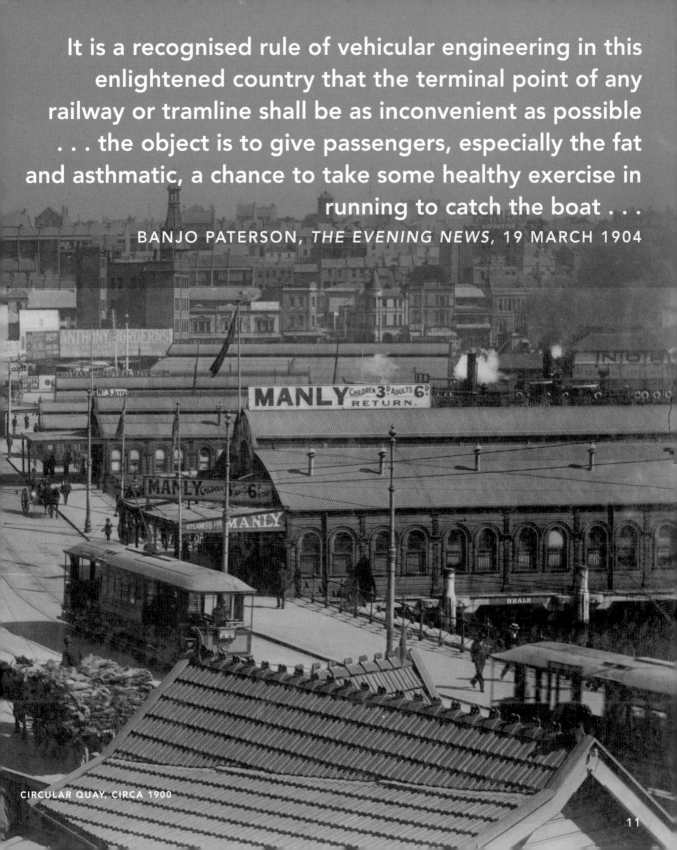

It is a recognised rule of vehicular engineering in this enlightened country that the terminal point of any railway or tramline shall be as inconvenient as possible . . . the object is to give passengers, especially the fat and asthmatic, a chance to take some healthy exercise in running to catch the boat . . .

BANJO PATERSON, *THE EVENING NEWS*, 19 MARCH 1904

CIRCULAR QUAY, CIRCA 1900

It was a slow and inefficient system of transport, made worse by the fact that few railway stations were close to ferry terminals. A journey across the city, perhaps only a few miles as the crow flies, might require catching a train, a tram, a ferry, another tram and then another train.

By the beginning of the twentieth century, Sydney was facing a crisis. The combination of a complex geography, poor or non-existent planning and continuous growth had created a city with water, sewerage and transport systems in danger of collapse.

A disaster finally forced the authorities to take action. In January 1900, bubonic plague broke out in The Rocks district and other inner-city slum areas. Hundreds were infected, 103 people died, and the lives of thousands were disrupted.

The outbreak threw into full public view the appalling living conditions in many of the older slum areas of Sydney, and it also exposed major inadequacies in the city's system of government, particularly in the areas of health, housing and planning.

The need to demolish and rebuild many of the old slums also made it possible to contemplate building a new transport system—including a harbour crossing.

They're shifting old North Sydney—
Perhaps 'tis just as well—
They're carting off the houses
Where the old folks used to dwell.
Where only ghosts inhabit
They lay the old shops low;
But the spirit of North Sydney,
It vanished long ago.

HENRY LAWSON

# Tunnels, bridges and endless talk

A tunnel beneath Sydney Harbour had been suggested as early as 1880. In 1892, a Royal Commission recommended that a bridge across the harbour be built once it was 'expedient' to do so. There was no shortage of plans put forward. In the first decade of the twentieth century, two separate Royal Commissions considered the merits of a bridge, or a tunnel, or a combination of both. The two commissions each recommended a different solution.

Parliamentary inquiries tried to settle the issue, with engineers, politicians and public servants arguing over the tunnel or bridge question, and what design should be used, and how to pay for such a costly project. At times one project or another seemed poised to proceed, but was scuppered by the political instability that plagued New South Wales.

The *Bulletin* joked that the harbour crossing was one of those things 'which are always coming to New South Wales, but which never arrive'. The bridge or tunnel had been 'coming for the last 30 years' but had 'broken down on the road every time'.

Closely following all these debates was an engineer working in the New South Wales Department of Public Works. J. J. C. Bradfield had first been employed by the department in 1891. Under the achingly slow system of promotion by seniority then prevailing in the public sector, Bradfield did not gain his first promotion to assistant engineer until 1909. But as early as 1903, Bradfield had discussed his interest in a bridge across the harbour. This was the project that would become his life's work.

At a parliamentary committee hearing in 1912, Bradfield gave evidence proposing a suspension bridge, but later submitted a design for a cantilever bridge.

In 1913, Bradfield was appointed Chief Engineer for Metropolitan Railway Construction and Sydney Harbour Bridge—even though the bridge and city railway were still not approved. In 1914, Bradfield was sent to investigate bridge and city railway systems overseas. He narrowly missed being trapped in Germany by the outbreak of World War I. The fighting put all plans for the harbour bridge on hold.

CABLES SUPPORTING THE BRIDGE ARCH
DURING CONSTRUCTION, CIRCA 1930

# Arches, beams and cables

BROOKLYN BRIDGE

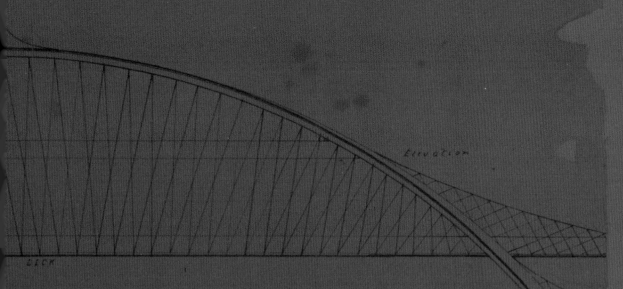

Throughout the nineteenth century, the availability of relatively cheap and well-made iron and steel allowed engineers to design bridges of a strength and span that previously would have been inconceivable.

Cable suspension bridges, like the Brooklyn Bridge in New York, used wire ropes anchored to tall towers to support the roadway. The other main styles of bridge used to span great distances were the arch and the cantilever, which was initially Bradfield's favoured design.

A cantilever is a beam that projects from a structure and is not supported at one end. A cantilever bridge uses two such beams stretching out from each shore, supporting a short middle section, which crosses the gap between them. The towers at each end of a cantilever bridge must support extremely high loads. The upper chords of the bridge are in tension; the lower chords are in compression. All this force is transmitted through the towers to the foundations.

In an arch bridge, the carriageway is held up by the vertical ties suspended from the arch. The entire load is transmitted through the arch to the base at each end, which in the case of the Sydney Harbour Bridge is formed by the four giant hinges. The towers at each end of the Sydney Harbour Bridge serve no structural purpose.

ANXIOUS FACES AWAIT THE RETURN
OF AUSTRALIAN SOLDIERS, 1919
AWM H11576

Future generations will judge our part in progressing
the nation by our works, and when designing the
Bridge, perhaps the largest structure ever erected
in Sydney, I have endeavoured to blend utility and
strength with beauty and simplicity so that the Bridge
can in some degree typify the resourcefulness and
idealism of our fallen men.

**WOUNDED AUSTRALIAN SOLDIERS, FRANCE AWM E00098**

# The last
# man and the last shilling

When World War I broke out, Australia joined the conflict cheerfully, enthusiastically. The young men who rushed to enlist could not have imagined the horrors ahead—no one did. The grim stalemate of trench warfare was not what any of the generals had planned. No one had intended for the slaughter and destruction to drag on for years.

Our national memory of World War I is dominated by the Gallipoli campaign, but it was on the killing grounds of the Western Front that Australia's most bloody and important battles were fought.

On the eve of war, Labor leader and future Prime Minister Andrew Fisher declared that the nation would fight for King and Empire 'to the last man and the last shilling', and Australia very nearly did so. Australia was on the winning side, but like Britain and France, victory had been achieved at fearful cost.

Some 360,000 men—this from a nation with a total population of four million—volunteered for military service. Of these, almost 60,000 were killed, and another 156,000 were wounded or captured. Almost one in ten Australian males of military age was killed.

The economic consequences of losing so many men in the prime of life were made worse by disrupted trade and the huge cost of the war effort, for which the Australian government borrowed heavily on the London market.

The social costs were also great. So many sons, brothers, fathers and husbands marched away and never returned. Their families did not even have bodies to bury. Many of those who did return were shattered in body and mind. Almost all of them found readjusting to civilian life extremely difficult.

The Australia to which the Diggers returned had changed dramatically. There was bitter division over the issue of conscription. Women had been allowed into jobs and held positions never before open to them. Price rises and shortages angered the working classes, who saw industrialists grow rich in war industries.

# Australia Unlimited

Overlaying these deep anxieties was a fragile gaiety, and a precarious prosperity based on borrowing and blind optimism. Writing in 1930, historian W. K. Hancock argued that Australian society was prone to alternating between naïve optimism and excessive cynicism. The period following World War I was one of foolish hopes:

> Men far away in comfortable cities dreamed of irrigation and agriculture in Australia's arid regions. The optimists began to preach, with the fervour of a tyrannical patriotism, their strange gospel of 'Australia Unlimited'.

This was the idea that, despite poverty of soil and rainfall, the Australian continent could somehow support a new United States, a world power with a population of perhaps 100 million people. 'Australia Unlimited' became both a widespread popular belief and a basis for serious policy.

J. J. C. Bradfield was one believer. He devised plans for the irrigation of large areas of the arid inland. With a vast complex of dams, pipelines and canals, he wrote, the Dead Heart could be made to bloom. From Alice Springs to Birdsville, the land would support sheep and cattle, rice and cotton, maize and wheat, tea and tropical fruits.

Bradfield's plan for the Sydney Harbour Bridge and the city rail system had much greater merit, but it too reflected the boosterism of Australia Unlimited. He used both the rhetoric of development and the memory of wartime sacrifice to build support for his plans.

'Future generations will judge our part in progressing the nation by our works', he said, adding that the bridge, 'perhaps the largest structure ever erected in Sydney', would 'in some degree typify the resourcefulness and idealism of our fallen men'.

Just as Australia's soldiers had proved themselves equal to any in the world, Bradfield's ambition was for a bridge that would signal Australia's emergence on the world stage. Like the Hoover Dam in the United States, it would be a symbol of confidence, progress and ingenuity.

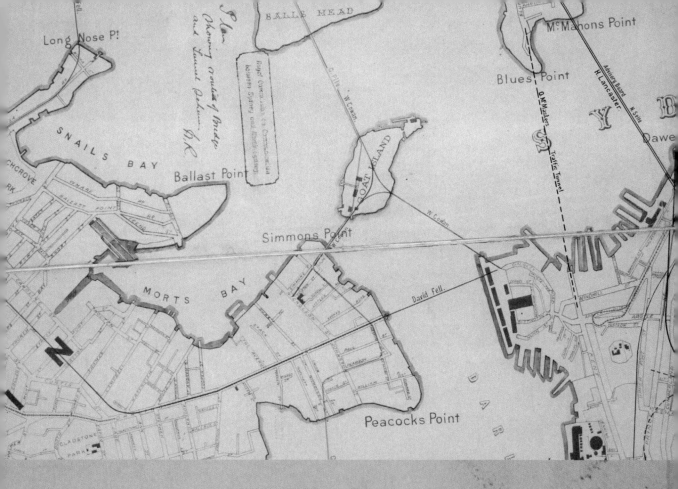

J H Cann
Minister for Public Works

Davis  5 Jul. 1913

Director General
of Public Works

J. J C Bradfield
Chief Engineer  26·4
Metropolitan Railway Construction

21

Do not think the Government is in earnest re the Bridge Bill. They have introduced it in such a way as to ensure defeat. However the work I have done will not be lost—it is one chapter in the romance of the Sydney Harbour Bridge, which someday I hope to write. Romance never dies; it is as beautiful as the line where the sun and sky meet beyond the boundless expanse of the oceans . . . Men of courage, unselfishness, endurance and patience win through and so it has been so it will be.

J. J. C. BRADFIELD FROM HIS DIARY

# The world's greatest engineer

Bradfield was an astute political operator. From the time of Federation, New South Wales endured frequent political instability. In the 1920s in particular, Labor and conservative administrations replaced each other from election to election, as if caught in a revolving door. Bradfield realised that this chaotic situation could be used to advantage. When governments are weak and short-lived, permanent civil servants rather than elected ministers can provide continuity of authority and policy—as long as they have the support of key politicians.

In 1920, Jack Lang became Treasurer in the Storey Labor government, and Bradfield approached him with his ambitious plans for an underground city railway and a road system centred around a harbour bridge.

Lang was impressed by Bradfield's vision and drive—uncharacteristic for public servants of the day—and agreed to support him. In and out of government, Lang backed the bridge and the city railway. Bradfield also cultivated members of the conservative Nationalist Party, but Lang was crucial. The Country Party was hostile to the project, which they saw as an extravagance that would benefit only the metropolis. The Country Party usually held the balance of power, and without the support of the Labor opposition, the *Bridge Act* would never have been passed.

Bradfield also astutely used the media to influence public opinion. He produced a seemingly endless stream of articles advocating his plan as a solution to Sydney's transport problems. He sometimes overdid it: in 1924 the *Daily Telegraph* called him 'the world's greatest engineer'. His high public profile made him enemies, particularly in the Railways Department, but combined with ceaseless lobbying, it enabled him to overcome the logjam of vested interests that made up the New South Wales government and Sydney's many municipal councils.

Throughout the early 1920s, Bradfield campaigned for his vision of a large bridge that would carry both motor vehicles and trains, and would be fully integrated into the city rail network. Eventually, on 16 November 1922, the *Bridge Act* was approved by the Legislative Council. Much was still uncertain: not least what the design for the bridge would be, and how the government would raise the money to pay for it.

It was not until Lang was elected Premier in 1925 that finance for the bridge was secured. Lang, who was also Treasurer, sent a high-powered delegation to London to raise the first in a series of loans. The amount of money was vast: the bridge would cost more than 10 million pounds, and the whole scheme, including the city railway, would come to more than 43 million pounds. The average weekly wage at the time was 4 pounds 5 shillings.

# The engineer's nightmare

On 28 July 1923, Bradfield turned the first sod for the beginning of construction of the bridge. The event was largely symbolic, staged for the photographers. Bradfield knew that public opinion demanded that, after all this delay, work 'get under way'. But even as he sank the ceremonial shovel into the earth, he was reconsidering the form the bridge should take.

He appeared for the cameras with a model of a cantilever bridge, the design that he had advocated at a public inquiry ten years earlier. But privately he had already changed his mind.

Bradfield feared that a cantilever bridge would be too heavy, too expensive and too risky.

A terrifying example of a failed cantilever bridge had stunned the world fifteen years earlier. Expensive and ambitious, the Quebec Bridge was an engineering marvel, the longest cantilever bridge in the world. It had been completed in 1917, but only after the structure collapsed twice in the course of construction. It was the stuff of an engineer's worst nightmare.

On 29 August 1907, two compression chords in the south arm of the bridge failed, and within seconds 19,000 tons of steel had collapsed into the river, killing 75 workers. A Royal Commission squarely blamed the project's chief engineers: 'The failure cannot be attributed directly to any cause other than errors in judgment on the part of these two engineers.' Work was resumed, but in 1916 the central section of the bridge fell into the river, killing eleven men.

Improvements in steel manufacturing had also made an arch design more attractive to Bradfield. A new silicon steel, lighter and stronger than any previously manufactured, was being made in the great steel smelters of England. It opened up new possibilities for bridge construction.

In New York, Bradfield saw and admired the Hell Gate Bridge, designed by Gustav Lindenthal, which had been built across the East River. Completed in 1916, the steel-arch span measured more than 1,000 feet. At a glance, Bradfield's architectural debt to Lindenthal is obvious, down to the unnecessary towers. Appearances, however, are somewhat deceptive. The Hell Gate Bridge carries only rail traffic, making it considerably lighter and avoiding many of the challenges posed by the twin-deck, mixed-traffic Sydney Harbour Bridge.

# 50,000 tons of steel

Bradfield's planned bridge was massive in every respect. By far the tallest structure in Sydney, indeed in the entire country, the bridge incorporated 50,000 tons of steel. It was by far the heaviest bridge yet built. The rivets used to hold the structure together were bigger than any previously used, and the steel angles were the largest that had ever been rolled.

Adding to the challenge, the depth of the water beneath the bridge meant that it would not be possible to support the arch while it was being built. Each arm of the arch would have to be held in tension by wire ropes as they crept out across the water.

The bridge's foundations were dug deep into the sandstone on each side of the harbour, and supported the four giant hinges that would eventually hold the entire weight of the bridge. Each bearing weighed more than 300 tons. Behind the hinges were U-shaped tunnels through which the anchor cables passed—128 for each side of the bridge.

The design required almost inconceivable precision. The arch was constructed from straight sections of steel, brought together at a slight angle. The maximum clearance allowed was $4/1000$ of an inch (0.1 mm) between angles. Bradfield was scrupulous in his inspections. In one instance more than 100,000 rivets had to be cut out and replaced because of a measurement discrepancy of $1/32$ of an inch (0.8 mm).

This cautious approach was vindicated. When the two halves of the arch eventually met, there was a combined error of only half an inch (13 mm).

The joining of the arch, on 19 August 1930, was a climactic moment. The remaining work, hanging the ties and installing the carriageway they would support, was relatively straightforward from an engineering point of view. The great arch, now locked securely in place, was a triumph for its creators.

However, even before its completion, a bitter row had erupted between Bradfield and his English colleague Ralph Freeman over whose arch it was.

There's blood upon the pylons; see
Pathetic on the stone,
A tablet for the eight men we
Remember died alone.

SUNG BY BRIDGE BUILDERS, PUBLISHED IN
THE UNION PAPER

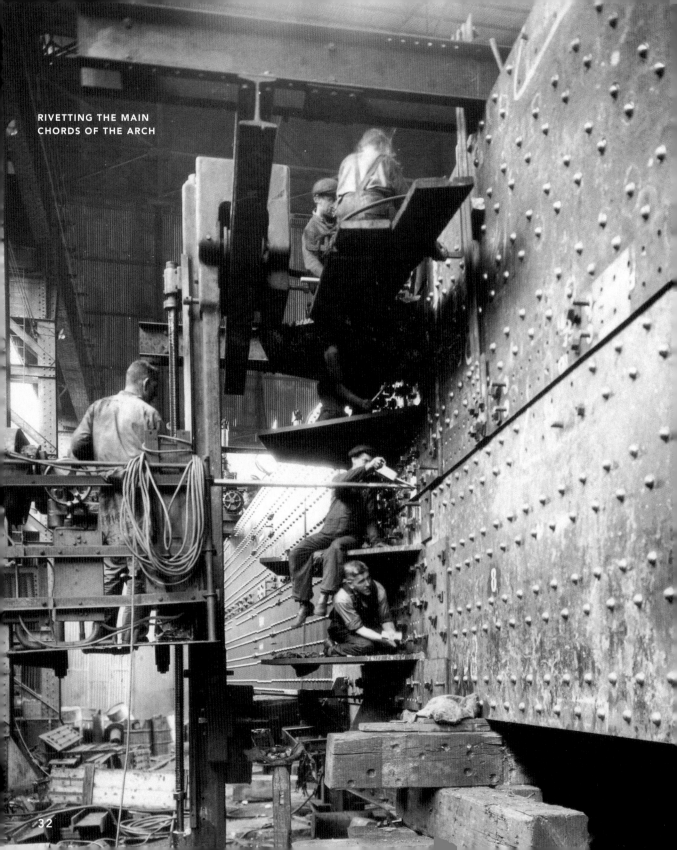

RIVETTING THE MAIN
CHORDS OF THE ARCH

It 'appened this way: I 'ad jist come down
After long years, to look at Sydney town,
An' 'struth! Was I knocked endways?
　　Fair su'prised?
I never dreamed! That arch that cut the skies!

C. J. DENNIS

# Alone I built it

In 1924, Bradfield had recommended that the tender to construct the Sydney Harbour Bridge be awarded to the great British engineering firm Dorman Long & Co. The company engaged the services of Ralph Freeman as a consulting engineer, and Freeman was in charge of designing the bridge, in accordance with Bradfield's specifications.

The two men worked well together for several years, but in 1928 Bradfield wrote an article about the bridge for *Engineering*, a prestigious professional journal. In the article, the bridge was said to be 'in accordance with the . . . design, plans and specifications [of] Dr J. J. C. Bradfield . . . The constructor's consulting engineers are Mr Ralph Freeman . . . and Mr G. C. Imbault'.

Freeman was furious. He wrote to Bradfield demanding that he be acknowledged as the designer of the bridge. Not only did Bradfield refuse, but from this point he claimed sole credit for the design. Plans and photographs of the bridge were marked 'Dr J. J. C. Bradfield . . . Chief Engineer'. Freeman was given no acknowledgment at all.

> *Dr Bradfield, reference to my connection with the Bridge in these publications is so rare that few people appreciate that I have any association of importance with the bridge. I have worked on the designs of the bridge for six years now. For the future may I ask that you use your influence to prevent publication of the statement that the 'design' of the bridge is yours?*
>
> RALPH FREEMAN

*Mr Freeman, I have to confess that I cannot understand your attitude. While I am specially anxious that all honour shall be paid where it is due, at the same time I feel that as the one who has been the most intimately associated with the project for many years the requests you make cannot be entertained.*

J. J. C. BRADFIELD

In 1929, Freeman hit back. *The Sydney Morning Herald* ran a series of articles crediting Freeman with the design and pointing to the similarities between the Sydney Harbour Bridge and the Newcastle–Gateshead Bridge in England, which had also been constructed by Dorman Long.

The matter was never satisfactorily resolved. When the bridge was opened in 1932, Dorman Long threatened to sue the New South Wales government if the official plaque described Bradfield as the designer. The press had great fun with the dispute. Cartoonists portrayed the two men, each pointing to the bridge and shrieking, 'Alone I did it!' But it was an unedifying squabble, and did neither Bradfield nor Freeman any credit.

Division continues to this day, on broadly nationalist lines. The *Oxford Companion to Australian History* says of the bridge: 'the design tender was won by railway engineer John Bradfield'. Freeman is not mentioned. The *Encyclopedia Britannica*, however, notes that 'in Sydney Harbour . . . Sir Ralph Freeman designed a steel arch bridge'. Bradfield is not mentioned.

In truth, the argument is one of semantics: What do you mean by 'designer'? The Bridge would never have happened without Bradfield's broad vision and political and managerial skills, and he set the project's parameters. But Freeman was responsible for many crucial elements of the design, and most of the detailed work. As one of Bradfield's colleagues observed of the bridge, 'It was big enough for both of them'.

> *Your specification for the Sydney arch is plainly reproduced from the outline of the Hell Gate arch. It is exactly the same type . . . Dr Bradfield has not inspired a single feature of the Sydney design.*
>
> RALPH FREEMAN

# The white city

Throughout the late 1920s, the arc of the great bridge slowly rose over Sydney. Long before its completion, the bridge dominated the city—much more than it does now, as there were few tall buildings to obscure it.

It was a work of breathtaking size and cost, a symbol of pride and progress, of Sydney's status as a world city.

By the end of 1929, the population of Sydney passed the one million mark. This made Sydney, after London, the largest 'white city' in the British Empire. It was a source of great civic pride.

For the first half of the twentieth century, whiteness was at the core of the Australian identity, a symbol not just of racial exclusion, but of a peculiar notion of purity and innocence. Australia was often personified in posters and on magazine covers as a young woman: cheerful, healthy and active, with white teeth and a slight tan, often wearing a white dress. She was innocent, pure, carefree.

If Australia was a girl in white, she led a very sheltered life. Immigration restrictions designed to preserve White Australia were just one facet of a culture of protectionism. She was protected from 'unfair' competition by tariffs, from exploitation at work by labour laws, from films that might upset or embarrass her by extensive censorship, from corrupting books by the *Customs Act*. Sydney, perched on the edge of Asia, was an astonishingly Anglo-Saxon city. And like the country as a whole, it tended to be hedonistic, inward looking, the result of what one commentator condemned as 'the policy of hermit Australia'.

But events that no law or government board could protect against were fomenting in the wider world. At the end of 1929, hermit Australia had its door kicked in.

43

. . . we were the most credit-addicted nation on Earth, and singled out for special punishment when the economies of the world hit the fan.

GERALD STONE

# The Great Depression

The economic disaster that swept the world in the early 1930s was without precedent. There had been financial collapses before, but nothing on such a broad international scale, nothing so severe, and nothing so apparently beyond solution.

What made the Great Depression so terrifying was that orthodox economics seemed incapable of explaining it. A Canberra journalist, Warren Denning, recalled that:

> Ministers, public servants, financial experts, economic advisers, bankers and businessmen were all equally adrift in a vast ocean of uncertainty.

This was not a recession due to a series of poor harvests, or the failure of some important industry, or the misconduct of a handful of bankers. Rather, the entire economic system built up by liberal capitalism, and which had transformed the world since the beginning of the nineteenth century, seemed to have collapsed.

Because of its high levels of public debt, the consequences were particularly bad for Australia. In the late 1920s, while most countries were worrying about the economic outlet and cutting back their borrowing, Australia sought loan after loan in London. In one period, the thirteen months from July 1927, Australian governments borrowed more than 63 million pounds: more than the market value of the entire wool clip.

Some bankers were alarmed. One report, by brokers S. R. Cooke and E. H. Davenport, said: 'In the whole of the British Empire there is no more voracious borrower than the Australian Commonwealth. Loan follows loan with disconcerting frequency . . . they get the money they want when they want it.'

By 1928, 28 per cent of Australia's export earnings were required to service debt repayments. The country was extremely vulnerable to an economic downturn. As the Depression began to bite internationally, commodity prices, particularly for the key exports wool and wheat, tumbled.

Australia's trade deficit for 1929–30 was 72 million pounds—a colossal figure for a nation of only 6 million people—and the interest due on existing loans was 23 million pounds. Overall national income fell from 645 million pounds in 1928–29, to 566 million in 1929–30 and down to 460 million in 1930–31: a contraction of some 30 per cent in two years.

12.6.2

9

Unemployment, which had already been high, soared. From about 10 per cent in 1929, recorded unemployment rose to 19 per cent by mid-1930, and to 28 per cent the following year. These figures recorded unemployment only among union members; the real rate was probably higher, reaching almost 33 per cent in 1932.

The Great Depression was like some biblical plague, striking down the just and unjust alike. The virtues so valued by middle-class Australia—thrift, prudence, hard work, honesty—proved no protection. Tens of thousands of people for whom self-reliance was a matter of deep pride found themselves bankrupt, unemployed and destitute. In hindsight, the fears people held at the time seem excessive, a collective hysteria, but this is unfair to those who lived through the Depression, who saw their world collapsing around them.

R. H. Milford, an ex-serviceman who travelled around Australia by car in 1932, was appalled by what he saw. 'Frequently, we would pass parties of young hoboes rambling the country looking for work. Some were abusive in their remarks called after us; but all looked pitifully inexperienced as "swagmen".' This, for Milford, was the writing on the wall. He worried about the vast ragged army of the unemployed and 'the terrible crop of criminals that must result from them within a generation'.

Anarchy or revolution seemed realistic prospects. There were, journalist Warren Denning recalled, rumours of riots, of an uprising, of an attack on Canberra by the unemployed: 'The air hung heavy with menace.'

R. H. Milford declared that:

*Something . . . must be done to prevent us finishing in the economic necropolis . . . and the present party-system must be thrown away—root and branch.*

Essington Lewis, the general manager of BHP, wrote that 'the ruling of the world by democracy seems to be an unmitigated disaster', and that a short-lived dictatorship 'would not be altogether a bad thing for Australia'.

There was much to criticise about the operation of parliamentary democracy in Australia in this period. The party system had rarely produced stable governments. The conservative side of politics was in disarray, unable to form a stable and coherent party. The Labor Party enjoyed some electoral success, but many of its leaders were inexperienced in government, and the party was prone to destructive factionalism.

Disillusionment and disgust with party politics caused a range of responses. One was the creation of the All For Australia League, a kind of anti-political political movement, which in time merged with the Nationalist Party to form the United Australia Party. Many rural regions wanted New South Wales to devolve into a group of smaller states, and formed separatist movements. Some conservative interests formed secretive groups, prepared to act as special constables in the event of civil collapse. The most spectacular and radical group, however, was the semi-fascist New Guard.

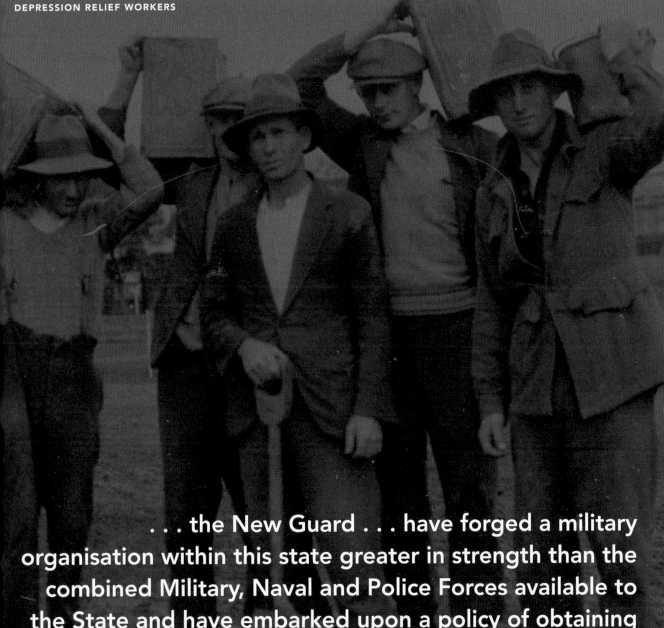

. . . the New Guard . . . have forged a military organisation within this state greater in strength than the combined Military, Naval and Police Forces available to the State and have embarked upon a policy of obtaining that organisation's desires by intimidation and force. If such an organisation is permitted to continue . . . as time passes, it will be a menace to this Realm.

POLICE REPORT ON THE NEW GUARD, 23 MAY 1932

We have no intention of handing over Australia to the tender mercies of the rubbishy kind of people who aspire to rule us . . . the best reply to force is greater force!

FRANCIS DE GROOT

# Fascism,
## with modifications

The New Guard was founded by a group of former army officers in February 1931. The Guard was organised on 'military principles', and aimed to abolish 'machine politics' and suppress 'any disloyal or immoral elements in Governmental, industrial and social circles'. Colonel Eric Campbell, a solicitor, emerged as leader. The movement grew rapidly as the economic and political crisis in New South Wales deepened.

One New Guard meeting was told:

*if the time comes when we have to defend our liberty God knows that we will be so organised that we can hit so hard and quick that we will succeed quickly . . . we have to strike quickly because it is only a well disciplined mobile force that will be able to achieve a quick victory which will prevent a lot of trouble. It would be murder if we are not half organised because by so being we will probably spill a lot of unnecessary blood.*

De Groot's famous ribbon cutting was the movement's high point. In April 1932 the New Guard began to splinter. In May, Jock Garden, a prominent unionist and key Lang supporter, was assaulted by members of the New Guard. This incident, and allegations of a New Guard plot to kidnap Lang and seize power, further weakened support for the movement.

That at least some elements of the New Guard seriously contemplated a coup is almost certain; whether it would have succeeded is less likely. The New South Wales Police, led by the colourful and unscrupulous Detective Chief Superintendent W. J. MacKay, ruthlessly suppressed the New Guard. After the incident at the bridge opening, MacKay told police dealing with a New Guard rally to 'go out there and belt their bloody heads off'.

Governor Game dismissed Lang and appointed Bertram Stevens, head of the newly formed United Australia Party, as caretaker Premier. After a bitter election campaign, Stevens won overwhelming endorsement at the polls in July. The New Guard faded from prominence at the same time, while becoming more openly fascist.

At its peak, in early 1932, the Guard had perhaps 80,000 active members; by November 1933, as few as 1,500 remained. For a time, though, the New Guard was a genuine threat to constitutional government in New South Wales.

# An unfair burden

The same people who conscripted our sons and laid them in Flanders' fields . . . now demand more blood, the interest on their lives . . .

JACK LANG

On the day in 1930 when Phar Lap won the Melbourne Cup, with 20,000 spectators watching over the fence because they couldn't afford an entry ticket, Jack Lang was sworn in for his second term as New South Wales Premier and Treasurer. His landslide election victory was, however, a most poisoned chalice.

The Depression caused bitterness and social division everywhere, but economic hardship and political turmoil were most serious in New South Wales. There were several reasons for this, but the most significant was the character of the government of New South Wales, and particularly its Premier, Jack Lang.

The problems that faced Lang on his return to office were immense. The economic disaster had drastically cut state revenue, while unemployment placed unprecedented strain on social services. Among Australian governments, New South Wales had been by far the worst offender in terms of reckless borrowing. The state government routinely funded public works by running up an overdraft, then later seeking a loan to cover the shortfall. As a result, the state's London bank accounts were usually in debit and loan repayments were increasingly difficult to meet.

As early as November 1930, Lang's private communications made the finances of the state government sound like those of a struggling corner store. 'My particular anxiety', he had written to the Federal Treasurer, 'is to find means of obtaining the necessary cash to carry on during the months of December and January'.

The Bank of England, one of Australia's main creditors, sent its representative, Sir Otto Neimeyer, to Australia to review the situation. Neimeyer's prescription was severe: a currency devaluation, cuts to award wages, balanced budgets and no further overseas borrowing until short-term debt was repaid.

A meeting of state premiers in May 1931 adopted a scheme along these lines. The Premiers' Plan, as it was called, was based on the principle of 'equality of sacrifice'. It is now commonplace to condemn this response to the Depression as counter-productive. By contracting economic activity, it is argued, the fiscal discipline imposed on Australia actually made the Depression worse. However, the premiers were following what the economic orthodoxy of the day dictated. It was, at least, a plan, and gave some promise of restored confidence.

Lang, on the other hand, was erratic and provocative. During the 1930 election campaign, he had rejected 'repudiation', that is, refusing to make foreign debt repayments. But in 1931 he announced that interest due to English banks would not be paid and the federal government was forced to meet the debt. Lang first agreed to, then renounced, the Premiers' Plan. By August 1931, the situation in New South Wales was desperate. After making some concessions to the other states, Lang was granted an emergency loan. Without it, the government would have been unable to pay the salaries of public servants.

Lang's rhetoric became increasingly extreme. At one public meeting he declared that the interest payment due 'was an unfair burden imposed upon the Australian people because of its participation in the war to help Britain'. This emotive claim was disingenuous on Lang's part. Only 16 per cent of Australia's interest payments were war-related. By far the bulk of the debt had been taken on in the past decade, some of it by Lang himself for projects such as the bridge.

# Whose bridge?

It was in this atmosphere of crisis that the bridge was completed, and its official opening approached. Bradfield oversaw an exhaustive program of testing the structure for strength and rigidity. One test involved parking 96 railway engines, with a combined weight of more than 7,500 tons, on the bridge at the same time.

The bridge itself was in fine shape, but tensions ran high over who would formally open it. Conservative opinion was shocked when Lang announced that he, rather than the Governor, Sir Philip Game, would perform the official opening.

In some respects, Lang's case was strong: this was a bridge paid for and largely built by the people of New South Wales, and he was the serving Premier and had played a major role in getting the project up. But in 1932, the British Empire and Australia's connection with and loyalty to the Royal Family were central to the nation's identity. A project like the bridge, constructed by a British firm using British steel and financed with loans from London, was also a symbol of Empire, and Australia's place within it.

The New Guard was most vocal in opposing Lang. 'The people of New South Wales will not permit Mr Lang to open the Bridge . . . we . . . will be forever dishonoured if we allow him who masquerades as Premier to open that bridge', declared New Guard leader Eric Campbell. There is evidence that some, at least, of the New Guard seriously intended a coup, or to kidnap Lang, in order to fulfill this vow.

Instead, Francis de Groot managed to gate-crash formalities.

In the name of the decent citizens of New South Wales, I declare this bridge open.

FRANCIS DE GROOT, NEW GUARD B SECTION COMMANDER, AFTER CUTTING THE RIBBON, 19 MARCH 1932

# The grand opening

There is no sound, and the picture is often blurred; even so, the film of New Guardsman Francis de Groot 'opening' the Sydney Harbour Bridge, on 19 March 1932, is dramatic.

De Groot, in military uniform and wearing a peaked cap, is on horseback. He raises his sword, and says something. Other sources record his words: 'In the name of the decent citizens of New South Wales, I declare this bridge open.'

A uniformed policeman grabs the horse's reins. De Groot says: 'I am a King's officer, stand back, don't interfere with me.' A dozen police officers stand a few paces away, but none moves.

Then the tall figure of Superintendent William John MacKay, officer in charge of the CIB and a future police commissioner, strides into frame. The camera swings to focus on the horse's head, but in the background MacKay can be seen forcefully pulling de Groot from his horse. De Groot lands in an untidy sprawl. MacKay hauls him up; the two men glare at each other, then de Groot is led away.

De Groot chose his stage well. The opening of the bridge was the largest public event ever held in Australia. More than one million people are said to have walked across the bridge on the first day. (One newspaper joked that they were real estate agents heading north to spawn.) There was a sense of celebration, of huge pride in the achievement the bridge represented.

Lang cemented this feeling by making a speech that emphasised the Bridge as a symbol of unity and nationhood:

*The achievement of this Bridge is symbolic of the things Australians strive for but have not yet achieved. Just as Sydney has completed this material bridge, which will unite her people, so will Australia ultimately perfect the bridge, which it commenced 30 years ago. The statesmen of that period set out to build a bridge of common understanding that would serve the whole of the people of our great continent. That bridge, unlike this, is still building.*

Sydney will be en fete today, and for many days, for a lavish programme of festivities has been arranged to celebrate the historic event . . . the scene in the crowded city today on the eve of the great day, was suggestive of the radiant sparkling atmosphere of Christmas and other festive periods.

*HOBART MERCURY*

NEY BRIDGE ELEBRATIONS BE THERE

D S ANNAN

# The dismissal

Lang's words were admirable, but ironic. As he spoke, his government was placing the Australian federation under tremendous strain.

In November 1931, six federal Labor members allied to Lang had, on Lang's instructions, voted against the Labor government of James Scullin, bringing it down. In the election that followed, the new United Australia Party, a moderate conservative coalition, headed by former Labor Treasurer Joe Lyons, came to power. Lyons passed the *Financial Agreements Enforcement Act*, often called the 'Garnishee Act'. This allowed the federal government to seize state revenues for debt repayment.

Lang responded by withdrawing more than one million pounds in cash from state government bank accounts. Public servants, escorted by police, carried the money in suitcases and stored it in the State Treasury. The Treasury locks were changed, and the building placed under a guard of unemployed unionists. Public servants were instructed to conduct all transactions in cash, and a circular was printed instructing them to disobey the Garnishee Act.

This last action, which was arguably illegal, was used by the Governor of New South Wales, Sir Philip Game, to justify dismissing Lang as Premier.

As historian Bede Nairn argues, Game's decision was constitutionally dubious 'but socially and politically Game was justified'.

Lang stood again in the July elections, but his sound defeat took the sting out of New South Wales politics. The New Guard rapidly disintegrated, and the threat of political violence faded.

Lang's successor in office, Bertram Stevens, has been called 'the forgotten Premier'. His obscurity is unfair, but easily explained. A colourless man, Stevens was the sort of leader historian Manning Clark derided as having the 'virtues of receivers in bankruptcy'. But there is something to be said for such virtues, particularly when the state is very nearly bankrupt. Stevens was honest and diligent, and over seven years as Premier he succeeded in restoring both the finances of New South Wales and confidence in the key institutions of state.

However, it is appropriate that Lang rather than Stevens opened the Sydney Harbour Bridge. Lang was a bold leader, a big-picture man, someone to whom Bradfield's ambitious visions appealed. Bertram Stevens could be relied on to run the Sydney Harbour Bridge efficiently and responsibly, but he would never have built it.

Jack Lang not only stood more than six feet tall, he threw his weight around. He knew what he wanted and he made sure that either he got it or nobody got anything.

GERALD STONE

Lang was a very big, imposing figure. And at a time when things were desperate, he needed to be a powerful advocate.

PAUL KEATING

GOVERNOR GAME (LEFT) AND LANG ARRIVE AT THE OPENING

# Icon or white elephant?

When the act enabling the construction of the Sydney Harbour Bridge was finally passed, long-serving Country Party member Colonel Michael Bruxner condemned it as a 'blot on the Statute book'.

Bruxner seems an unlikely critic of the bridge. A responsible and diligent politician, he later served as transport minister and made important contributions to the development of infrastructure across New South Wales. But Bruxner was expressing a view that was widely held in rural New South Wales: the bridge was too expensive and represented an unjustifiable investment in a metropolis that was already too dominant in the affairs of the state.

To criticise the Sydney Harbour Bridge now seems like shooting Skippy: unthinkable, if not un-Australian. The bridge remains a vital part of Sydney's transport infrastructure. Without the bridge and the city railway, Sydney as we know it could not function, because Sydney as we know it is the result of Bradfield's great plan.

Perhaps even more important, the bridge is our most distinctive man-made icon, the centrepiece of many a fireworks display and Qantas advertisement, not to mention millions of kitsch souvenirs.

But perhaps Bruxner had a point.

Sydney is too big. Its population has grown beyond its natural limits. Its roads are choked, its public transport infrastructure under strain. The water supply is inadequate, and sewage treatment and disposal presents endless headaches. Housing is ridiculously expensive. The dominance that Sydney exerts in the politics and economy, not just of New South Wales but of the whole of Australia, is often unhealthy.

If the bridge had not been built, it would have retarded the development of Sydney—and perhaps that would not be such a bad thing. Had the bridge not been built, Sydney today would be a smaller and more liveable city. Other capital cities and other centres in New South Wales would be larger, our economy, government and culture less centralised.

But the bridge *was* built, and to this day it is an inspiring structure.

The visitor to Sydney cannot fail to be impressed, standing by the heaving waters of the harbour, looking up at this vast arch: awesome in its scale, elegant in its geometry, beautiful in its setting. Bradfield said that 'Future generations will judge our part in progressing the nation by our works'. Seventy-five years later, nothing that Australians have built quite matches the great bridge of which he dreamed.

# Pipe Dreams

The dream: a man-made river through the desert, which would unlock vast riches, transform an isolated colony into an economic power, and make Federation possible.

# The players

## C. Y. O'Connor

Charles Yelverton O'Connor was born in Ireland in 1843. The O'Connors were landed gentry, part of the Protestant Ascendency that then ruled Ireland—but unusual members of it. When the great famine gripped Ireland in the years after Charles' birth, his father John took the biblical injunction to 'go, sell your possessions and give money to the poor' literally. John O'Connor gave up all his resources, eventually having to sell his house and estate, to provide food relief for starving families. The family was forced to move to a modest house in Waterford, and young Charles spent most of his childhood living with relatives. His father's ideals of selflessness and public service would mark C. Y. O'Connor all his days.

O'Connor broke with family tradition to take up what was then the new profession of engineering. As a teenager he showed a staggering capacity for hard work and great ability as a surveyor and financial manager. At the age of twenty-one he migrated to New Zealand, where he found work surveying for the construction of roads and railways, and assisting in the construction of a permanent harbour at Westport. By 1883 he had been appointed New Zealand's under-secretary for public works.

In 1890, he was recruited to work as Chief Engineer in the newly self-governing colony of Western Australia. There he formed a productive partnership with the colony's larger-than-life Premier, John Forrest. O'Connor reorganised and extended the railways, and his bold and innovative design for Fremantle Harbour transformed it from a shallow roadway, suitable only for small ships, to a

# John Forrest

Born near Bunbury, Western Australia, in 1847, John Forrest was one of ten children of Scottish immigrants. In 1869, Forrest was chosen to lead an expedition in search of Ludwig Leichhardt. The party found no trace of the missing explorer, but Forrest led his party through some 2,000 miles of uncharted wilderness, systematically surveying and collecting scientific specimens as he went.

In 1870, Forrest led another expedition, the first to cross from Perth to Adelaide by land. In 1874 he set out again, crossing from Geraldton to Mount Peake, and turning south to Adelaide. This epic journey, towards the end of which he came upon the overland telegraph wire, made Forrest a celebrity.

Appointed to the West Australian Legislative Council in 1883, Forrest rose rapidly to dominate the politics of the colony. Despite his humble background, his diligent administration and straightforward manner won him the strong personal allegiance of many people. When Western Australia achieved self-government in 1890, he was the natural choice to serve as Premier.

Forrest was scrupulously honest and careful in his use of public money. Equally, he was determined to drag Western Australia from its position as poor cousin among the colonies. He believed this required large-scale public works, funded by borrowing from abroad.

With some difficulty, he led the colony into Federation in 1901, and then played a prominent role in the new Commonwealth Government. He died in 1917, while sailing to England to seek treatment for cancer.

major international port. However, it is the Gold-fields Water Scheme, the system of pipelines that made the major gold centres of Kalgoorlie and Coolgardie habitable, for which O'Connor is best remembered. O'Connor was as imaginative and forward-looking as he was meticulous in the attention to detail required by his profession. Universally known as 'the Chief', he was witty, good company, and a devoted family man. Scrupulously honest and fair-minded himself, he was easily wounded by the malice of others. This tendency to be thin-skinned, combined with the stress and exhaustion of managing the pipeline scheme, is believed to have led to his tragic death by his own hand in 1902.

# The Cinderella state

The European settlement of Western Australia began with a failed experiment. The Swan River Colony, as it was initially known, was the first colony to be established as a free settlement rather than as a penal colony. It was sponsored by a syndicate, which put up capital for the project in return for land grants. But the scheme had little chance of prospering. The first group of settlers to arrive, in 1829, had been drawn by reports of rich grazing land, but they were inexperienced in farming in Australian conditions, and had inadequate capital. Worse, the glowing descriptions of fertile land proved false. There were few areas of good soil, and these were widely scattered. The summers were long and hot. The indigenous people, understandably, were not pleased at the new arrivals, and hostility and violence resulted. The free settlers were hamstrung by the lack of an adequate labour force.

But the major problem facing the settlers was a simple fact about the Australian environment: water is scarce. Ours is the driest inhabited continent on Earth. Aridity is its predominant characteristic. Two-thirds of Australia is classed as either desert or semi-arid. Even those areas that enjoy better rains are subject to seasonal dry periods and longer climatic cycles that bring severe drought.

European settlers struggled to adapt to this demanding environment. In some ways this struggle continues. The Australian environment is profoundly different from that which shaped our forebears' ways of living and thinking. Even the word 'drought', which implies that a dry season is an aberrant event, reflects a culture and language that evolved on damp Atlantic islands.

The West Australian colony survived, but only just and as a stagnant backwater.

By 1848, the settler population numbered only 4,600. In desperation, the people of the colony petitioned for convicts to be sent there, hoping to use their labour on roads and other public works.

Transportation did stimulate the economy, but this dubious benefit ended with the last shipment of convicts in 1868. Exploration, particularly that conducted by the Forrest brothers, led to the discovery of areas of good grazing land, and wheat farming began to enjoy some success. Still, in 1888, the *Bulletin* would caricature Western Australia as 'the Cinderella state', a tag that struck for decades.

But under the energetic leadership of John Forrest, the colony was already beginning to shake off its lethargy. After winning the right to self-government in 1890, Forrest was free to borrow money from abroad to finance the bold plan of public works that he believed was necessary. He just needed to find an engineer to build them.

SWAN RIVER COLONY WAS FORMED BY SETTLERS DRAWN BY GLOWING REPORTS OF FERTILE LAND

What is the object of building railways and opening up the country? That the country and people here may flourish and develop the resources and turn the desert into the smiling plain, and build up a great country and make it a worthy offshoot of the great Empire whence we or our fathers came.

JOHN FORREST

FREMANTLE HARBOUR

# Railways, harbours, everything

In March 1891, a convention to discuss the possibility of Federation was held in Sydney. Premiers and senior ministers from every Australian colony and New Zealand spent six weeks debating the issue, without deciding very much. But Forrest asked around about possible engineers, and C. Y. O'Connor, then working for the New Zealand government, was recommended to him.

Political interference in public works appointments in New Zealand had made O'Connor increasingly unhappy with his position, and Western Australia offered enormous opportunity. When O'Connor inquired what his duties would be, Forrest cabled this reply: 'Railways, harbours, everything'.

It was an inspiring call, but in pursuing his employer's ambitions for developing Western Australia, O'Connor quickly made enemies.

Politics in nineteenth-century colonies were conducted with a viciousness and personal spite that makes the most strident modern talkback radio seem tame. Civil engineers, responsible as they were for expensive public works projects that could make or destroy personal fortunes, were particular targets.

O'Connor was determined to reorganise the operation of the railways: the amount of track in the colony was small, and machine tools and rolling stock were few and out of date. But his drive for efficiency and modernisation upset existing interests, accustomed to the slow pace and bureaucratic restraints that had prevailed before self-government.

His innovative plan for Fremantle Harbour, too, required all O'Connor's tenacity to succeed. When work on the harbour began in November 1892, John Forrest praised his chief engineer's vision and commitment. Forrest had, he conceded, at first been opposed to O'Connor's plan, as he feared it would cost too much. But O'Connor 'stuck to his scheme; he urged it with all his power', and Forrest, along with parliament, was persuaded.

> In this action of the engineer you see the character of the man; he was not afraid to take the responsibility of this great work. I believe that we have in him an able and energetic, a brave and self-reliant man, and I only hope in this great work he has undertaken that he will be successful.

O' Connor lived up to Forrest's hopes. Within three years, smaller ships were able to enter the river and use the new docks that had been built. In May 1897 the SS *Sultan,* of more than 1,200 tons displacement, steamed proudly up the river; in 1899 the *Barbarossa* (10,700 tons) did the same. Fremantle had become a major international port and the natural first port of call for all ships approaching Australia from the west. To this day, the harbour is vital to the West Australian economy.

# The magic metal

Gold, more than any other single factor, transformed the Australian colonies.

Before gold, the local elite had aristocratic pretensions. They envisaged an orderly, thinly populated land, in which graziers would build beautiful houses on large estates. The labouring classes, bonded or free, would be the new tenants to a new gentry.

Gold ruined this pastoral idyll. The timing helped, quite by accident. In 1851, New South Wales and Victoria, followed by Western Australia in 1890, were granted self-government just before major gold discoveries were made. Gold drew new populations in such numbers that the old colonists were swamped.

The diggings—chaotic, ill-serviced, rough places—were the unintended crucible of Australian democracy and egalitarianism. Since ancient times gold has held the promise of crashing through social barriers. Folklore is full of stories in which poor young men, at their last extremity, are helped by some magical creature to find buried treasure. The hero becomes wealthy beyond measure and marries a princess.

The real diggers, of course, did not always find much gold—and often squandered what they did find—but they were men and women who aspired to change and improvement. The communities they formed were not easily governed, and the wealth they found gave them at least a taste for independence.

When the surface gold gave out, the diggers often took up some new means of making a living. But the rough democracy that gold inspired stayed with them, and shaped the nation's civic traditions and forms of government.

By the 1890s, the heady days of the rushes in eastern Australia were long gone, but the dream returned on the far side of the continent. In the late 1880s, alluvial gold was found in sites scattered across Western Australia, but only in small amounts. Then, in July 1892, while resting their horses on a grassy flat the Aboriginal people called Coolgardie, two prospectors found gold literally poking through the soil. In an hour, they had picked up 20 ounces of gold in nuggets. The rush was on.

Some 4,500 miners arrived in 1892; 5,000 the following year. In June 1893, a few days' ride from Coolgardie, two prospectors found more surface gold: in two days they collected more than 100 ounces. Kalgoorlie, as the place became known, was the top of an enormous gold reef.

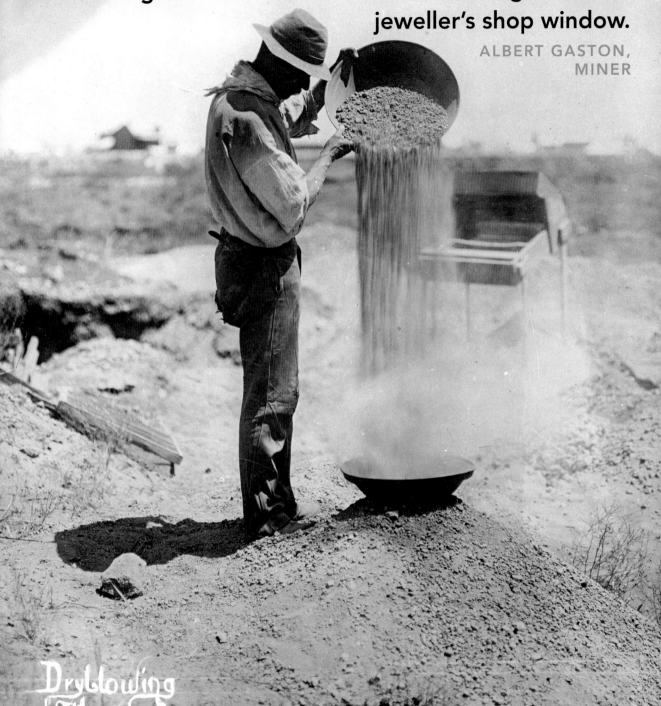

I enquired my way to Bayley's claim and there I saw the first coarse gold I had ever seen. It was shining out like a jeweller's shop window.

ALBERT GASTON, MINER

Dryblowing with Dishes 172 Dwyer Photo. Kalgoorlie

The lure of spectacular riches triggered a staggering influx of migrants and money. The non-Aboriginal population doubled—to 101,000 between 1890 and 1895—and kept soaring. There were around 160,000 by 1897.

Many of the new arrivals came from Victoria and South Australia: 't'othersiders', the locals called them. In smaller numbers came gold seekers from England, the United States (including a future president, Herbert Hoover) and China. The need for desert transport brought another exotic group—camel drivers from Afghanistan. With the local Aboriginal population, they formed a rough and colourful new community.

The pickings were rich. By 1896, the gold yield was more than 280,000 ounces, worth more than 1 million pounds; by 1902 the value of the yield was nearly 8 million pounds. British investors—wary that the other Australian colonies were mired in recession—poured millions of pounds into West Australian mining ventures.

It was soon clear that the Kalgoorlie region would sustain a considerable mining industry for a long time—or rather it could, if one major obstacle could be overcome.

A LETTER FROM HOME

This was the last great water
famine in Kalgoorlie . . . men
were walking about with their
lips parched and cracked; too
thirsty to speak.

ALBERT GASTON, MINER

# A dry, rich land

The goldfields at Coolgardie and Kalgoorlie are located on a vast plain, most of it some 1,400 feet (420 metres) above sea level. The rainfall is only about 5 inches (13 centimetres) per year. There are no rivers. The only lakes are salt. Aboriginal people had sustained themselves in the region for millennia, but in small numbers, moving frequently. They knew where water could be obtained at particular times of year. Of particular importance to them were *gnamma*, natural holes in rock that captured rain water.

If mining were to succeed, extra sources of water were needed. The government sank bores to tap artesian supplies, but the only water found was brackish. It had to be desalinated to be fit to drink. This was done using condensers, in which the water was boiled, and the steam captured and cooled. But the output of potable water was literally a dribble.

Water was limited to drinking and to the most minimal hygiene.

The gold itself had to be won from the earth not by sluicing but by 'dry blowing', pouring alluvial soil from a height so the wind would 'winnow' the lighter soil and leave the gold. This was oppressive work. 'Surely no form of labour is more exasperating than that of dry-blowing,' wrote one miner. 'Dust gets into the eyes, clogs the nose, and makes the throat as dry as a lime kiln.'

More seriously, the shortage of water meant that sanitation in the camps and towns of the goldfields was appallingly inadequate. Visitors complained of the putrid smell and thousands of flies. Human waste contaminated the soil and dust, which then settled on the roofs. Every time it rained, this dust contaminated the water tanks and typhoid would stalk the goldfields.

As the population of the mining towns soared, so did the number of peopled killed by typhoid fever. In 1892, there were 55 deaths from the disease. Three years later the number had risen to 325; the year after that to 400. The toll did not begin to decrease until 1898. In all, nearly 1,900 people are known to have died from typhoid in the 1890s; the true figure is probably higher.

The colonial government tried to create adequate water supplies, building dams that could hold millions of gallons, but these reservoirs were hopelessly outstripped by the population, growing at tens of thousands per year. The dams were, in any case, often distant from the mining areas, and the water had to be transported by camel train, selling at its destination for as much as 5 shillings a gallon (about 4 litres).

By 1895, John Forrest, who had been reluctant to invest in infrastructure for the region in case the rush proved short-lived, was now convinced that an adequate water supply must be provided. He turned for a solution to his engineer-in-chief, the man who had done so well on the Fremantle Harbour, C. Y. O'Connor.

There is no doubt we have treated the blacks very badly. We have taken their country from them, and destroyed their game . . . We have taught them all the white man's vices and then left them to die of starvation and disease. It is to our lasting shame.

ALBERT GASTON, MINER

# Where the white man finds gold the white man stays

The sudden arrival of large numbers of men and animals into the parched and delicate land around Kalgoorlie was catastrophic for the indigenous people. Clara Saunders, one of the few women who travelled to the goldfields in their early years, told how an Aboriginal woman refused to show her party where gold could be found.

*'No, no,' she went on to say. 'No show'um gold'. If she did she would have a curse put upon her [by her tribe] . . . the blacks know that where the white man finds gold the white man always stays and the black man has to go.*

The boom towns that sprang up across the goldfields voraciously consumed the sparse open forest that covered the plain. Wood was needed for mining props, and especially for fuel. Unfortunately, the salmon gum was poor fire-wood, burning hot but quickly. 'The countryside has been devastated of timber for miles around,' noted one miner. This woodland had been an important hunting ground and source of plant foods for the native people.

Perhaps the worst impact was on the region's most scarce resource, water. Clara Saunders lamented how, as the miners made water a commodity, it was effectively stolen from the land's traditional owners.

# Rivers in
# the desert

The problems confronting O'Connor were immense. The only area where there was adequate rainfall to fill a dam, and where a dam could feasibly be built, was in the Darling Ranges, near the coast and more than 300 miles from the goldfields. But the distance, though great, was not the main problem.

Between the ranges and the goldfields is a long, gentle, but substantial rise. The water's destination was more than 1,000 vertical feet (300 metres) above any dam that could be constructed.

Long pipelines had been built elsewhere in the world, but one running uphill over such a distance had never been attempted. Pumps driven by steam engines were used in many cities to lift water into storage reservoirs, but only over short distances.

O'Connor and his staff at the Public Works Department developed thirty-one different proposals, of which three were presented to Parliament.

The scheme finally chosen was audacious and expensive. A storage reservoir would be constructed in the Darling Ranges. Water would be taken from it by 30-inch diameter steel pipes. Eight pumping stations, each with service tanks and a holding reservoir, would push the water over 329 miles to Kalgoorlie. This pipe would deliver 5 million gallons of water per day. Each pumping station would lift the water about 225 feet.

**MUNDARING WEIR, DARLING RANGES**

The expense, even for a colony flush with gold, would be enormous: 2,500,000 pounds. It was a lot to borrow to fund an engineering project that had no precedent anywhere in the world, and which might not work.

A particularly difficult decision was what sort of pipe should be used. At the time, most large water-supply pipelines were made of cast iron. It was a strong material, long lasting and dependable. However, it performed less well under high pressure, particularly if the pipes were big.

Steel pipes were becoming more common. Steel was lighter than cast iron, which also made it cheaper to transport, and it could resist greater water pressure. Steel did have its drawbacks, however. It was more likely to corrode, and it was not possible at the time to make large steel pipes out of one piece of metal: it was necessary to join separate sheets with a watertight seal.

The cheapest and most common way to do this was to use rivets. Riveted pipes were likely to rust, however, and the bumpy surface of the pipes disturbed the flow of the water. Over a short distance, these tiny ripples were of little significance—but over hundreds of miles, they greatly increased the friction that the pumps had to overcome.

O'Connor eventually decided to use a new sort of pipe made with what was called a locking bar. Two semi-circular sheets of steel were used, their edges being joined by two long bars. The edges of the sheets were slightly thickened, and the locking bar squeezed tight around them: much like a long, continuous dovetail joint. The locking bar pipes were more expensive, but because the inside of the pipe was smoother, there was less friction for the pumps to overcome. The locking bar design was also stronger and more reliable than rivets. This was important: the water in the pipes would exert pressures of up to 400 pounds per square inch (more than ten times that exerted by the air in a car tyre).

The research into the many technical questions the pipeline posed, the surveying for a suitable dam site, and for the hundreds of miles of pipe track, all took time. When the pipeline proposal was put to Parliament in July 1896, John Forrest grandly quoted the prophet Isaiah: 'They made a way in the wilderness and rivers in the desert.' But as time passed, the mining communities were rancorous about the delay, and an increasing number of critics both condemned the vast expense of the scheme and began to question if it would work at all. When the construction of the main dam, at Mundaring in the Darling Ranges, was seriously delayed, one newspaper commented: 'The progress being made . . . seems to indicate that the first half-pint of tarry fluid will reach the goldfields somewhere about AD 2000.'

The time being taken to complete the scheme was becoming part of a major political storm: whether Western Australia would join the other colonies in a new nation. The colony was close to tearing itself apart.

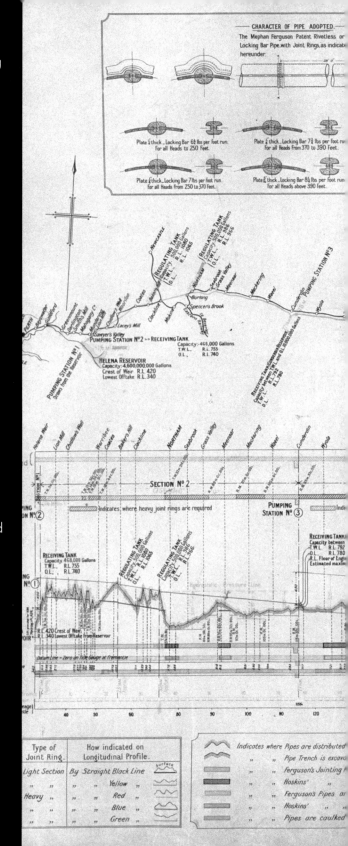

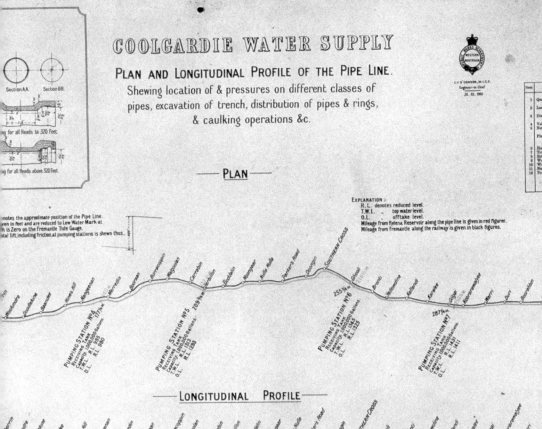

# One dissoluble Federal Commonwealth

Novelist and republican Tom Keneally once observed that the Australian Constitution is 'as dull a document as the Articles of Association of the Dee Why Bowling Club'. It is hard to quarrel with this assessment, but the document does contain a few clues to the ferocious political struggles over the shape of the new nation, which continued to the eleventh hour.

For example, the preamble declares that 'the people of New South Wales, Victoria, South Australia, Queensland and Tasmania . . . have decided to unite in one indissoluble Federal Commonwealth'.

Where, you might wonder, is Western Australia? It does not get a mention until clause 4, which says that 'if Her Majesty is satisfied that the people of Western Australia have agreed thereto', then that colony will become part of the new nation.

Section 26 of the Constitution sets out the number of members for the House of Representatives in the first federal Parliament, state by state. But there are two tables: one has Western Australia as a state, the other does not.

These oddities are a permanent reminder that Western Australia squeezed through the doors of the Federation train as it was pulling out of the station. It was, almost literally, a last-minute decision.

There were several reasons for Western Australia being the last and most reluctant colony to commit to Federation. Its sheer remoteness set it apart from the other colonies, the vast reaches of desert cutting it off as surely as a thousand miles of ocean. Western Australia's slow development was part of the picture, too. Economically weak, and struggling to attract migrants, the colony had only just won the right to self-government from England. There were fears that Federation might mean exchanging rule from distant London for rule from distant Melbourne.

Gold changed the picture: not only was Western Australia economically much stronger, but the miners, many of them migrants from the east, were far more enthusiastic about Federation than the older settlers. This division created a political nightmare for John Forrest.

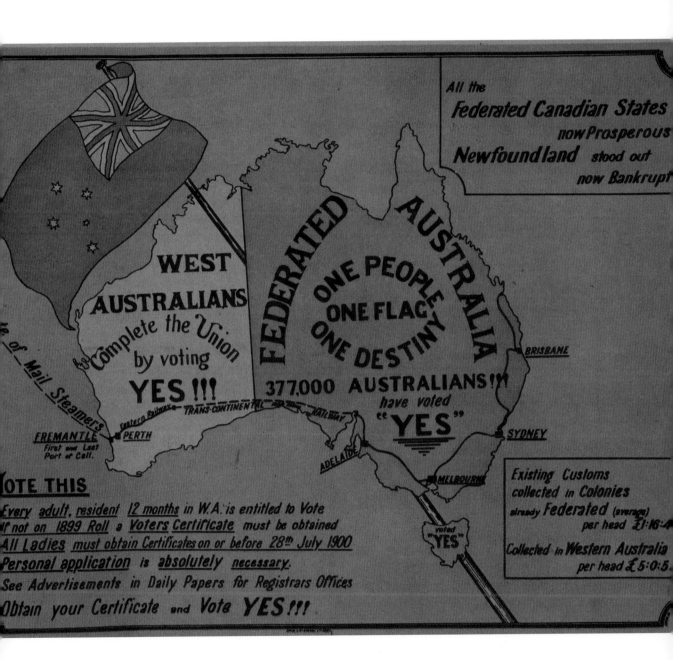

All the **Federated Canadian States** now Prosperous **Newfoundland** stood out now Bankrupt

**WEST AUSTRALIANS** Complete the Union by voting **YES !!!**

**FEDERATED AUSTRALIA**

ONE PEOPLE ONE FLAG ONE DESTINY

**377,000 AUSTRALIANS !!!** have voted **"YES"**

*e of Mail Steamers*

*Eastern Railway* **TRANS-CONTINENTAL** *Railway*

**FREMANTLE** First and Last Port of Call.

**PERTH**

**ADELAIDE**

**MELBOURNE**

**SYDNEY**

**BRISBANE**

voted "YES"

**OTE THIS**

Every <u>adult</u>, <u>resident</u> 12 months in W.A. is entitled to Vote
f not on 1899 Roll a <u>Voters Certificate</u> must be obtained
<u>All Ladies</u> must obtain Certificates on or before 28th July 1900
<u>Personal application</u> is <u>absolutely necessary</u>.
See Advertisements in Daily Papers for Registrars Offices

Obtain your Certificate and Vote **YES !!!**

Existing Customs collected in Colonies already **Federated** (average) per head £1:16:4

Collected in Western Australia per head £5:0:5.

93

# Land of gold

Some sort of federation of the Australian colonies had been suggested as early as 1846. But progress was agonisingly slow: meetings, conferences, councils and consultations came and went. The colonies often agreed in principle to the desirability of Federation, but found the devil in the detail. Federation seemed likely in the early 1890s, but foundered because of the reluctance of New South Wales. As the nineteenth century drew to a close, however, an agreement again seemed achievable.

Forrest, a supporter of Federation for almost the whole of his public life, was placed in an extremely delicate political position. The core of Forrest's political support came from the settled farming areas, but it was these regions where the greatest apprehension about Federation lay.

Forrest needed to extract the best possible deal for rural interests if Western Australia as a whole were to join. But the population of the goldfields, dominated by 't'othersiders', was enthusiastic about Federation, and saw Forrest's caution as the dithering of a Perth-centred squattocracy.

Forrest was already unpopular among the miners, who blamed his government, often unfairly, for the failure to provide services to the goldfields. Attempts to restrict alluvial miners from sinking deep shafts caused serious riots. Forrest himself was almost assaulted at Kalgoorlie in March 1898.

If Australia could speak with one voice, how much more important would she be? If her tariffs were identical what a market within herself for free competition would there be . . . If we are to become a nation, to be the great power in the southern hemisphere, it can only be by being federated.

JOHN FORREST

HANNAN STREET, KALGOORLIE

"The Tide's Rising: Which way will I jump?"

Speaking for the colony I represent, I say that amongst the electors there is no great feeling in favour of federation. It will take a good deal of persuasion . . .

JOHN FORREST

By 1900, the other five colonies had agreed to Federation. Forrest had been unable to secure support in the state Upper House for a popular referendum, and as the deadline for joining the new nation approached, the gold-fields began to talk of taking matters into their own hands. Their proposal was to create a separate colony, 'Aurelia', Latin for Land of Gold, which would then unite with the rest of Australia.

On Thursday afternoon I saw accused among the crowd he called out where is Forrest mob the bugger kill the bugger he called out several times, afterwards I saw him on a Lorry he was addressing a Crowd—and said boys we will have our rights if Forrest does not give it we will give him another Eureka Stockade.

POLICE EVIDENCE AGAINST A MAN ACCUSED OF INCITING THE 1898 KALGOORLIE RIOTS

Unless the Eastern Goldfields can be separated from Western Australia, there will be in this liberty-loving part of the world the curious anomaly of a large community forced to accept and obey an oligarchy that robs them of their rights and justly receives their contempt and detestation.

*KALGOORLIE MINER, 1898*

If separation succeeded, Western Australia would be left as a rump colony, huge in area but small in population, and with an economy based largely on agriculture. That produce would face quarantine and customs restrictions imposed by the rest of Australia. Worse, the colony would have a huge debt burden, the legacy of Forrest's public works program, but no access to mining revenue to service it.

Forrest's position was weak, and he knew it. The other colonies did not need Western Australia, and they refused to grant Forrest extra concessions for his colony.

But Forrest was a skilled politician. At the eleventh hour he was able to persuade his reluctant supporters that Western Australia should join the Federation as a foundation state, or risk being forced in on worse terms later. A new bill for a referendum was approved in May 1900. Six weeks later, on 31 July, the people of Western Australia overwhelmingly voted in favour of joining the new nation of Australia, which would come into being just five months later.

Having dextrously manoeuvred his reluctant colony into Federation, Forrest then threw his considerable bulk—he weighed almost 20 stone (127 kilograms)—into making the new Commonwealth government a reality.

Elected unopposed to the Federal Parliament, he departed for the interim capital, Melbourne. But if Federation for Western Australia was a personal triumph for Forrest, and the gateway to fresh challenges, it was a disaster for his protégé, the engineer who had done so much to help realise Forrest's vision for Western Australia.

No government has the right to ask us to sanction expenditure on a scheme which is bristling with falsehoods, the figures of which are grossly misleading and deductions there from are false.

DEBATE ON COOLGARDIE GOLDFIELDS WATER SUPPLY LOAN BILL, 1896

All that O'Connor knows about engineering could, without crowding, be stated in a very small book . . . .
*SUNDAY TIMES*, 22 OCTOBER 1899

This man has exhibited such gross blundering . . . he has robbed the taxpayers of millions of dollars.

JOURNALIST

# Joining the pipes

The pipes used to construct the Goldfields Pipeline were each 28 feet long. This meant that more than 100,000 joins had to be made to complete the pipeline, each one watertight under pressure and tough enough to resist the extreme changes in temperature experienced in the desert.

The pipes were joined by inserting the ends into an 8-inch steel ring, or sleeve. This was then packed with hemp, lead and tar, a process called caulking. Traditionally, this was done by hand, which was time-consuming and labour-intensive, and came at the risk of damaging the pipes.

A Victorian engineer, James Couston, invented a machine to caulk the pipe joins. Although it eventually worked well, Couston's machine was complex and suffered many teething problems. It took time for workers to learn how to operate the new machine, and many hand-caulkers resented being displaced from their jobs.

Amid mounting criticism about the time being taken to complete the pipeline, Couston suggested that his company would be able to complete the job more quickly if it was contracted to complete the work. O'Connor agreed, and recommended to the minister that Couston's firm be given the work. O'Connor also suggested that tenders not be called, as time was pressing and Couston's firm was the only possible choice.

This decision, sensible from an engineering point of view, was politically disastrous. It allowed O'Connor's critics to accuse him of corruption.

At the end of January 1902, the *Sunday Times*, long a fierce critic of O'Connor, began a new assault over the caulking machine and the contracts awarded to its inventor, James Couston.

# The position has become impossible

With his patron no longer Premier, and occupied with national affairs in distant Melbourne, the sensitive O'Connor faced a vicious campaign of vilification. It would drive him to his death.

By the early summer of 1902, O'Connor was showing obvious signs of strain and depression. Normally a witty and talkative man, he would lapse into sudden silences. The final push seems to have been the announcement, in February, of a Royal Commission to 'Enquire into and Report upon the Conduct and Completion of the Coolgardie Water Scheme'.

The inquiry was the result of accusations of corruption being made in state Parliament and the press. There were two major areas of complaint. One was O'Connor's decision to award, without a tender, a contract to a private company to finish the remainder of the pipeline. The other was some questionable land dealings involving T. C. Hodgson, the project's second-in-command.

The decision to contract out some of the construction of the pipeline followed a bitter controversy over the Fremantle Harbour project. O'Connor usually preferred that large projects be controlled directly by the government, rather than contracted out to private firms. Potential contractors were consequently hostile, and took every opportunity to attack the workmanship and cost of O'Connor's projects.

O'Connor was repeatedly forced to defend himself and his department, at length and in detail. He stressed that he was not wedded to using departmental labour, and preferred to use contractors where this was appropriate. The intemperate language of the *Sunday Times* was matched by debates in the state Parliament. The wildest innuendo and smear were bandied about the chamber.

Since Federation, and Sir John Forrest's departure for Melbourne, West Australian politics had been dogged by instability. This was partly Forrest's legacy: he had never formed a political party, but commanded a strong personal following. In his absence, a series of short-lived and weakly led ministries came and went.

It is open rumour everywhere that this shire engineer from New Zealand has absolutely flourished on palm grease since the first day when the harbour works and the Coolgardie Water Scheme were agreed upon. If he is not now immensely rich there is some mystery somewhere. And apart from the distinct charge of corruption this man has exhibited such gross blundering or something worse in his management of great public works that it is no exaggeration to say that he has robbed the tax payer of many millions of money . . .

*SUNDAY TIMES*, 9 FEBRUARY 1902

When the Couston controversy exploded, the Premier was George Leake, who had opposed the pipeline at its inception. The Minister for Works, C. H. Rason, had been in the portfolio for only a few weeks. Neither man offered more than a lukewarm defence of their engineer-in-chief.

On 19 February, a Royal Commission into the Goldfields Pipeline was announced. It was clear O'Connor himself would be the main target of the inquiry. He would have to again defend and justify his every decision, with little sympathy or support from the government. O'Connor became visibly worried and distressed.

In the early hours of the morning of Monday, 10 March, O'Connor sat at his desk and wrote his suicide note.

**My father was so absorbed that he hardly seemed to realise the change of country although I am sure there were times when he felt the needs of old friends. A loneliness then indescribable, it must have been.**

KATHLEEN O'CONNOR,
*MEMOIR OF HER FATHER*

*The position has become impossible. Anxious important work to do and three commissions of enquiry to attend to. We may not have done as well as possible in the past but we will necessarily be too hampered to do well in the imminent future.*

*I feel that my brain is suffering and I am in great fear of what effect all this worry may have upon me—I have lost control of my thoughts.*

*The Coolgardie scheme is all right and I could finish it if I got a chance and protection from misrepresentation but there's no hope for that now and it's better that it should be given to some entirely new man to do who will be untrammelled by prior responsibility.*

*Put the wing walls to the Helena Weir at once*

Exhibit. 8 R

The position has become impossible

Anxious important work to do and three commissions of enquiry to attend to

~~If we have got emphatic over business in the past we~~

We may not have done as well as possible in the past but we will necessarily be too hampered to do well in the immediate future

I feel that my brain is suffering and I am in great fear of what effect all this worry may have upon me — I have lost control of my thoughts

The Coolgardie scheme is all right and I could finish it if I got a chance and protection from misrepresentation but theres no hope for that now and its better that it should be given to some entirely new man to do who will be untrammelled by prior responsibility

10/3/02

But the wingwalls to Helena Weir at once —

> No honourable, strong minded man is afraid of public criticism. It is only men of weak intellect, or men who feel and realise some oppressive guilt upon their consciences who are perturbed by suggestions upon their conduct . . .

*SUNDAY TIMES*, ON THE SUICIDE OF C.Y. O'CONNOR, 16 MARCH 1902

At about 6 a.m., O'Connor came out of his house, walked towards the stables, and asked a stableman to ready his horse. There was nothing unusual in this: O'Connor often rode out early in the morning. Usually, his daughter Bridget accompanied him, but on this day she was feeling unwell. About half an hour later, O'Connor reappeared, and after briefly returning to the house, saying he had forgotten something, he mounted his horse and rode off towards Fremantle's South Beach.

There were no witnesses to what followed. According to the evidence given later at a coronial inquest, O'Connor dismounted, letting the horse go free. Standing in about two feet of water, facing towards the land, he took out his false teeth and put them in his pocket. He placed the muzzle of a revolver to his mouth and pulled the trigger.

O'Connor's tragic and unnecessary death caused a wave of public grief—exacerbated among some, perhaps, by a sense of guilt. John Forrest, in Melbourne when he learnt of O'Connor's death, said the tragedy had left a 'gap which cannot now be filled . . . I mourn with the people of Western Australia the loss of one who has left behind a high and honourable record of splendid public service, and I mourn the loss, also, of a dear and valued friend'. Thousands attended the funeral; thousands more lined the cortege route.

If the *Sunday Times* felt any remorse, it was hard to pick. The paper merely protested that its criticisms had always been motivated by the public interest, and dropped the grubby hint that O'Connor's suicide might reflect a guilty conscience.

If the malice directed at O'Connor was not refuted by the successful completion of the Goldfields Pipeline, just ten months after his death, it must surely have been by the winding-up of his estate.

Far from prospering through corrupt deals, O'Connor had owned no land. His personal property consisted of two horses and a cow, some furniture, books and other personal effects. The Public Works Department owed him 35 pounds in unpaid wages. After various small debts were paid, his entire estate was valued at 189 pounds, 5 shillings and 10 pence. If not for a life insurance policy, which provided a modest income, his family would have been left almost destitute.

The coroner who inquired into O'Connor's death found that he had committed suicide 'while in a state of mental derangement caused through worry and overwork'. Modern psychiatry would use more technical, and perhaps more tactful, language, but the basic finding cannot be questioned.

O'Connor's suicide is sometimes discussed as if it were mysterious: Why would he kill himself when all the serious problems with the pipeline had been solved, and successful completion of a great scheme was within sight? But there is no great puzzle. O'Connor was a classic example of the driven high-achiever: a person with high personal ethical standards and a strong—perhaps excessive—work ethic. People of this character are vulnerable to depressive illness, and in the weeks before his death O'Connor exhibited many of the symptoms of serious depression.

This was entirely understandable. He was exhausted, worn down by seemingly endless personal attacks, and felt isolated and unsupported by his political masters. To be faced suddenly with the need to justify himself, yet again, before an inquiry likely to be hostile to him must have seemed an unendurable burden to a weary man. Shakespeare knew the yearning: 'To die: to sleep . . . 'tis a consummation devoutly to be wish'd.'

But understanding O'Connor's death does not make it any less tragic. When, on 24 January 1903, on an oppressively hot day, Sir John Forrest turned the wheel that opened the valve that sent the water streaming into the Mt Charlotte Reservoir, near Kalgoorlie, C. Y. O'Connor was a palpable absence.

O'Connor's biographer, A. G. Evans, describes the moment eloquently:

*His very absence on that occasion seems to haunt us like a spectre in a historic photograph . . . His absence at Kalgoorlie on that day—that blank space—is an uncomfortable reminder of the cruelty of circumstances, the fickleness of public opinion, and our helplessness in the face of others' personal sufferings.*

I pay tribute to the memory of O'Connor, the great builder of this work . . . I am greatly saddened that he did not live to receive the honour so justly due to him.

**JOHN FORREST**

# Legacy

The Goldfields Pipeline, the main sections of which were completed in 1903, was one of the engineering marvels of its time, a source of great national pride. Never before had water been moved so far through a pipe, let alone uphill.

There was a magic, an aura, to the very vastness and audacity of the pipeline. It brought water to the desert—not just any desert, but a place where, beneath the parched earth, lies one of the richest gold deposits in the world, one of very few that has been continuously mined for more than a century.

In some respects, the pipeline broke new ground. The locking bar steel pipe, the invention of Victorian engineering firm Mephan Ferguson, was an ingenious new design. James Couston's caulking machine, for all the trouble it caused, was also a significant innovation.

But O'Connor used, so far as possible, tested technology and orthodox design. The pumps used were a standard British make, and by breaking the pumping into eight stages, the loads and pressures were kept within prudent limits.

It is ironic that O'Connor's critics thought his plan wildly impractical, and those praising it later thought it incredibly daring. The genius of O'Connor's scheme, to the contrary, lay in its careful pragmatism. He was, to his bones, an engineer. He was a practical man, meticulous in his attention to detail. He had a dream, and set about realising it methodically, responsibly, and using the disciplines of his profession. He built his pipeline only when he was convinced that it would work.

In the decades after his death, C. Y. O'Connor became the stuff of legend. He was presented in numerous school texts as a tragic hero, first of the Empire and later, as times changed, of Australia.

Such mythologising has a cost. Geoffrey Blainey is one historian who has warned against the dangers of 'sanctifying' O'Connor. In his history of Kalgoorlie, Blainey argues that the pipeline, while a brilliant feat of engineering, 'was not necessarily an object lesson in the massive use of public funds'. Only the fact that the Kalgoorlie mines turned out to be richer by far than was anticipated at the time of construction saved the pipeline from being an economic failure.

But while Blainey is right to reinforce the need for dispassionate historical inquiry, C. Y. O'Connor remains an extraordinarily appealing historical figure. His dignity, scrupulous honesty and ethic of public service were remarkable for his time. A century later, living in the age of spin, these qualities shine all the more brightly.

The most telling tribute to O'Connor comes from an unlikely source: economist and historian Edward Shann.

Shann, best remembered for his classic economic history of Australia, published in 1930, was scathing of the tendency of Australian governments to borrow heavily to finance grandiose public works. 'They were building,' he said of them, 'in haste and on credit, the nineteenth century equivalent of city walls.'

But Shann made an exception for John Forrest, whom he acknowledged as a leader who was bold but not reckless, who would borrow to build, but only what was worth building.

Shann also praised C. Y. O'Connor's integrity, diligence and skill: 'It was his [O'Connor's] brain that enabled Forrest, by public works built with British capital, to remove the great natural impediments to colonization in the south-west corner of the continent.' O'Connor had been destroyed, Shann lamented, by the very people who owed most to him.

**OPENING OF THE PIPE, 1903**

Future generations, I am certain will think of us and bless us for our far-seeing patriotism, and it will be said of us as Isaiah said of old, 'They made a way in the wilderness and rivers in the desert.'

JOHN FORREST

# A Wire through the Heart

We live in a connected world.

When something big happens, good or bad, we know about it, almost instantly.

There is so much communication, it can be too much.

Sometimes we yearn for some peace and quiet. For empty spaces in our days, and time to think. For the world before the wire. But that world was dominated by distance, by isolation, by the agony of not knowing.

In the nineteenth century, it took as long as five months for a letter to arrive from London, a ten-month round trip.

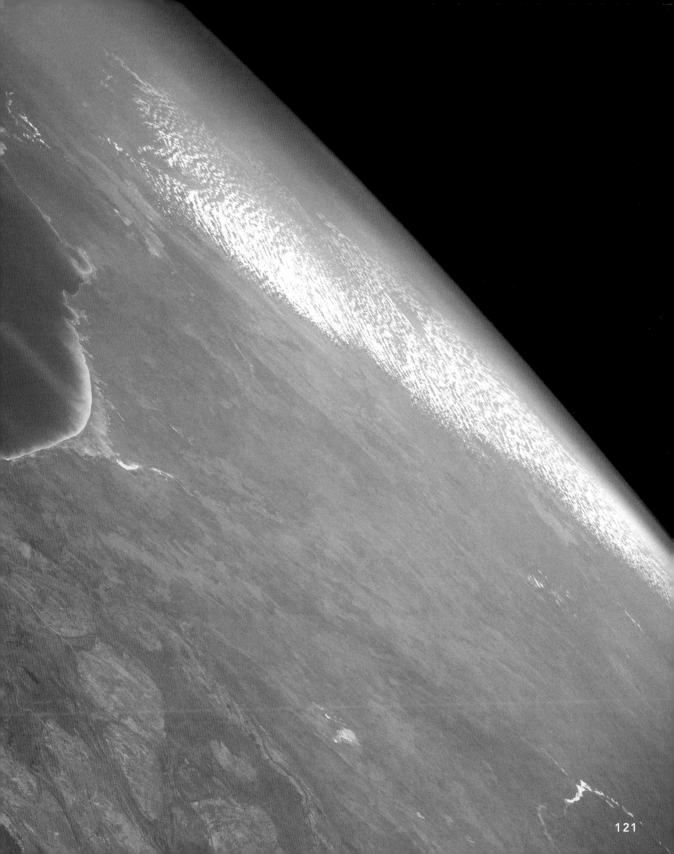

## John McDouall Stuart

John McDouall Stuart was born in 1815 in Fife, Scotland. He attended a military academy before deciding, while still in his teens, to migrate to South Australia. After working in surveying, he accepted an offer from the great explorer Charles Sturt to join an expedition to the interior of Australia in 1844. Sturt's expedition was a seventeen-month odyssey that tested the endurance of the party to the limit, but it gave Stuart an unrivalled knowledge of the topography and the difficulties of travelling in the arid centre.

For the next twelve years, Stuart worked as a surveyor and real estate agent, but never lost his ambition to return to the interior, vast areas of which were still unknown to Europeans. In 1858, he embarked on the first of five epic journeys, eventually reaching the Timor Sea on 24 July 1862. However, his health was shattered and he was nearly blind. He returned to Scotland in 1864, then moved to London.

A prickly, solitary man, Stuart was fiercely independent, and was disliked by some influential people in South Australia. He had a reputation as a heavy drinker, and alcoholism may have contributed to the brain haemorrhage that caused his death in London at the age of fifty.

# The players

# Charles Todd

The son of a grocer and tea merchant, Charles Todd was born in London in 1826. His ability in mathematics and astronomy gained him a junior position at the Greenwich Observatory in 1841. The observatory was experimenting with the use of the electric telegraph to transmit time signals, and Todd became fascinated by the potential of this new apparatus. In 1855, when the government of South Australia decided to employ a superintendent of telegraphy, Todd was recommended for the position.

Chiefly remembered for his role in constructing the remarkable Overland Telegraph, Todd also made an important contribution to connecting Western Australia to the telegraph system in the 1870s, and to the development of scientific meteorology in Australia.

A cheerful and optimistic man, Todd was a skilled negotiator. Todd's engineering achievements were recognised both in Australia and in Britain, and he was knighted in 1893.

Todd married Alice Bell shortly before leaving for Australia. Their relationship was a long and happy one. They had six children of their own, and also raised three other children left orphaned when Todd's brother died. They were prominent supporters of the Congregational Church. Alice Todd, after whom Alice Springs was named, died in 1898; Charles Todd died twelve years later in 1910.

# Richard MacDonnell

Richard MacDonnell was an Irish-born lawyer who was appointed Governor of South Australia in 1855, just as the colony was poised to achieve self-government. He had previously served in vice-regal posts in Africa and the Caribbean. A brash, authoritarian figure, he was ill-suited to negotiating the transition to self-government and alienated many liberals. However, MacDonnell was energetic and interested in exploring. He encouraged and supported Stuart in his expeditions, and is widely credited with having been the first to propose an overland telegraph line from Darwin to Adelaide.

MacDonnell left Adelaide in 1862, and served as governor in places as far afield as Hong Kong and Nova Scotia before retiring.

# A tale of
## two princesses

On 31 August 1997, Diana, Princess of Wales, was killed in a car crash in the Pont de l'Alma tunnel in Paris. Her driver was travelling at terrific speed, trying to elude pursuing media photographers. Diana's death was made known almost instantly around the world, and in many nations, including Australia, an almost endless stream of broadcasts, news specials and tributes flowed in print, over broadcast media and on the internet. Her funeral, on 6 September, was broadcast live and watched by an estimated audience of 1 billion people.

On 5 November 1817, Charlotte, Princess of Wales, died during childbirth in Surrey, England. Only twenty-one, Charlotte had been married little more than a year. After two miscarriages, her third pregnancy went to term, but after fifty hours of labour she delivered a stillborn baby boy. A few hours later, Charlotte died herself, probably as the result of an internal haemorrhage.

The tragedy shocked the British people, and there was public mourning across the nation and across the Empire.

The young colony of New South Wales was not left unaffected: on learning the news, the newspapers rushed to print special editions, bordered in black. Sermons were given in memory of the late princess. Monuments to honour her were planned.

The depth of public feeling was similar to the global experience after the death of Diana 170 years later, but with one exception.

The news of Charlotte's passing had to travel by mail coach and packet ship, halfway across the world. The people of Sydney did not read about the event until 2 April 1818. By the time the officers of the New South Wales Corps were making a formal resolution of sympathy and mourning, Charlotte and her stillborn child had been buried for five months.

The colonies that were to become Australia were settled in an age when travel and communication were slow, costly and unreliable, even in much of Western Europe. News travelled, as historian Tom Standish puts it, 'outwards in a slowly growing circle, like a ripple in a pond, whose edge moved no faster than a galloping horse or a swift sailing ship'.

The isolation experienced by the European population of the early Australian colonies was almost absolute. The convicts and their gaolers were achingly remote from their homes, families and friends, not just in space, but in time. News of a battle, or a birth, or a wedding would take five months—often longer—to arrive. A reply would take as long to return.

Historian Geoffrey Blainey argues that Australian society has been shaped by what he dubbed the tyranny of distance. Not only were the Australian colonies dauntingly remote from their mother country, they were also remote from each other.

The problems faced by all settler societies—establishing viable agriculture, maintaining law and order, providing basic services from banking to medicine, maintaining efficient and accountable administration—were made more difficult by the vast distances within and between Australia's colonies.

The tyranny of distance has been proffered as a reason why Australians enthusiastically adopt new communications technologies. We love to keep in touch. In recent years, we have taken to mobile phones and text messaging with the same gusto as we did to the electric telegraph when it first became available, 150 years ago.

# Farewell to Old England

Many folksongs of the time record the anguish of people long separated from loved ones whose fate is unknown. In one song, 'There was a Lady', a young woman sings of her lost love:

> It's seven years since I've had a sweetheart
> And seven years since I did him see
> But seven more I will wait upon him
> For if he's alive, he'll come home to me

This particular song ends happily: the lady's sailor returns, having made his fortune. In reality, most of those transported to Australia, and many of those who went of their own free will, never returned to the land of their birth. The song 'Bound for Botany Bay', now a cheery campfire song, was then a real lament. Many of those who sang it really *were* saying 'Farewell to Old England forever'.

# What hath God wrought!

The first large-scale telegraph systems in Australia were based on the technology of an American, Samuel F. B. Morse. Born in Massachusetts in 1791, the eldest son of a clergyman, Morse's first ambition was to be a portrait artist. But at Yale College, along with his art studies, he attended lectures on chemistry and electricity, and was fascinated. He wrote to his family about one experiment in which the students all joined hands, those at each end then touching a battery terminal: 'the whole class taking hold of hands and form the circuit of communication and we receive the shock at the same moment'.

Electricity was one of the great mysteries of the age. Its allure is illustrated by Mary Shelley's *Frankenstein*: it is electricity that is used to give life to the monster. Electricity was known, but understood in only the barest ways. That it had extraordinary qualities was obvious. But how it worked, how it could be harnessed, what might not be achieved with it—these were tantalising questions.

Despite years of struggle, Morse never quite succeeded as a portrait painter. His interest in electricity, however, led him to develop a system of communication that would change the world. Using electricity and wires—first joining cities, then spanning continents and crossing oceans—the Morse telegraph would turn the world into a great 'circuit of communication'.

Morse completed the first telegraph line in the United States, from Baltimore to Washington, in 1844, and sent the first message over it: 'What hath God wrought!'

The Morse telegraph was the foundation of the astoundingly connected world in which we now live. And when something terrible happens—perhaps as personal as the death of a loved one, perhaps as universal as the terrorist attack on the United States in 2001—like Morse's classmates at Yale, 'we receive the shock at the same moment'.

# Morse code

A junior secondary student could easily make a simple electric telegraph. You connect a wire to a battery. At the far end, you connect the wire to something that shows that current is flowing—a light bulb, say—and then earth it. At the battery end, attach a simple switch. You can use a switch to connect and then break the electrical circuit; almost instantaneously, the light bulb will go on and off. Devise a code that is understood by the person watching the light bulb, and you can readily communicate complex messages over long distances.

But this device seems simple to us now only because we have a sound theoretical understanding of electricity, and because the necessary materials are easily obtained. The pioneers of electrical engineering had neither the knowledge nor the hardware. They discovered what are now basic facts about electricity—for example, that a small amount of current at a high voltage will travel much further along a wire than a large current at a low voltage—through trial and plenty of error. And equipment available for a few dollars in any hardware shop these days, such as reliable batteries and insulated conducting wire, were difficult to manufacture.

It was in this context—a pioneer science fumbling its way forward, testing new ideas and slowly building from experience—that the genius of the Morse telegraph must be appreciated.

Earlier systems of telegraphy had failed or were ineffective in part because of their complexity. One system, for example, used a set of six wires that made a bank of three needles point to a pattern of letters. With primitive equipment and an imperfect understanding of how electricity worked, such a complex system had endless teething troubles.

The Morse telegraph used a code based on breaking and connecting the current along a single earthed wire. The code had only two elements: a short burst of current, the 'dot', and a long burst, the 'dash'. It was simple, flexible, and could be understood even if the transmission signal was poor.

The code Morse first devised in 1835 was improved in 1851 to allow languages other than English to be readily transmitted. Apart from some minor changes made in 1938, the International Morse Code remained unchanged for the whole of the telegraph era. It is still used in radio communication.

. . . a telegraph extended to join the several seats of commerce in Australia and also, it is no idle dream in the present age of wonders . . . connecting [to] Asia via submarine cable then via Calcutta to London.

CHARLES TODD

# A scheme vast and difficult

It took nine years, and the discovery of gold, for the colonies of Australia to show interest in the electric telegraph. The pioneer of the technology in Australia was Samuel McGowan, an Irish-Canadian who had studied under Morse. Inspired by the fortunes being made on the gold diggings, he arrived in Melbourne with some telegraphic instruments and batteries.

McGowan persuaded the Victorian government to support him, and he built the first telegraph line in Australia, covering the short distance between Melbourne and Williamstown. Within three years, Melbourne was connected to Ballarat, Wodonga and Portland.

Though expensive, the telegraph service was extremely popular, and the government of South Australia saw the potential for Adelaide to become the information centre of the colonies. In 1855, they sought out an expert in telegraphy to develop a system for the colony, and Charles Todd was recommended. With his new wife, Alice, he arrived in Adelaide that year.

Todd quickly completed a short telegraph line between the city and the port, but his reputation was made when he supervised the construction of the South Australian section, some 300 miles (480 kilometres), of the Adelaide to Melbourne telegraph.

Already Todd was thinking big. As early as 1855 he wrote of his vision of a telegraph 'extended to join the several seats of commerce in Australia and also, it is no idle dream in the present age of wonders . . . connecting [to] Asia via submarine cable then via Calcutta to London'.

There is some dispute about who first conceived the idea of running an overland telegraph from Adelaide to Darwin. Todd later claimed that he was the originator, but Sir Richard MacDonnell, Governor of South Australia and an enthusiast for exploration, probably deserved more credit. It was MacDonnell who encouraged and supported Stuart in his explorations, and as early as 1858 approached the British Treasury for finance to build an overland telegraph.

Todd was also keen to link Australia to the outside world. 'Our ultimate telegraph communications, via India, with England,' he wrote, 'a scheme vast and difficult . . . will, we doubt not, at no very distant date be carried out.' However, Todd favoured a completely different route. The international link from Java would come ashore at Port Essington, in what is now the Northern Territory, then run via a long submarine cable around Cape York and down the Queensland coast to Moreton Bay.

Todd's objection to an overland route was simple: no one knew what lay in the forbidding interior of the continent. Was it possible to cross—never mind build a telegraph through—this vast and unknown country?

The man who provided the answer was John McDouall Stuart.

# The wee Scot

Stuart was a man of remarkable tenacity, a gifted surveyor and perhaps Australia's greatest explorer. He was also solitary, aloof, and had a drinking problem. This alienated many in South Australia, which, alone among the Australian colonies, had been established entirely by free settlers, and which prided itself on its religious probity.

When, in 1858, Stuart began his five epic journeys of exploration into the interior of Australia, the region was as mysterious as it was forbidding to the white settlers. South Australia was faced with a barrier of salt lakes, especially the massive Lake Eyre, which isolated the colony. Many thought that a shallow inland sea, salty and devoid of life, must lie beyond.

Stuart was able to cross the lakes, and in his first expedition in 1858 he discovered vast tracts of potential grazing land, though much of it was marginal at best. With under-resourced expeditions and scarce food and water supplies, malnutrition and illness repeatedly forced him and his party to return.

On his third expedition he reached the geographical centre of Australia, now named Central Mount Stuart, and came within 200 miles (320 kilometres) of country in the north of the continent that had already been explored. In 1860 he led an attempt to cross the continent, south to north, a race with the Victorian party led by Burke and Wills. Stuart was unable to cross Sturts Plain, and was forced to return to Adelaide, but in October 1861 he set off again. This time, despite Stuart's failing health, his party finally reached the Timor Sea on 24 July 1862.

He had crossed the entire continent, from south to north, passing through the centre of Australia, a distance of more than 1,800 miles (2,900 kilometres). The Victorian expedition led by Robert O'Hara Burke had reached the Gulf of Carpentaria sixteen months earlier, but Burke and all but one of the party who made the final dash to the sea perished on the return journey.

Stuart was very nearly subject to the same fate. Suffering from scurvy and nearly blind, he managed to return to Adelaide, only the second person to cross the continent and return alive.

He was awarded 3,000 pounds in recognition of his achievements, but was only allowed access to the interest, not the principal—a reflection of official concern about his intemperance.

Bitter at his treatment, fighting dementia and alcoholism, Stuart returned to Britain in 1864, where he died two years later.

Despite his personal limitations, Stuart's achievements were extraordinary. By crossing the continent he had shown that there was no impassable inland sea, that there was enough water to support men and animals, and that there were Aboriginal people who managed to live off the apparently barren land. In his published journals he suggested that a telegraph line could be successfully built through the interior.

It was Stuart's success in crossing the continent that persuaded Todd to change his view about the best route for a telegraph. He came to believe firmly that an overland line was the better option.

133

All that remained was to build—in eighteen months—a telegraph wire from coast to coast, passing through the arid heart of Australia. The journey alone had nearly killed Burke and his party eight years earlier.

N

W    E

S

Darwin
Yam Creek
*July 1862*

Katherine

Daly Waters

*May 1861*

Powell Creek

NORTHERN
TERRITORY

Tennant Creek
*June 1860*

Barrow Creek
*May 1860*

Alice Springs

Charlotte Waters

SOUTH
AUSTRALIA

*Jan 1860*

The Peake    *Lake Eyre*

Strangways Springs

Beltana    *Lake Frome*

*Lake Torrens*

Port Augusta

Adelaide

0   100   200   300   400km

——— Telegraph wire

Stuart's explorations
- - - 3rd expedition
- - - 4th expedition
- - - 5th expedition
- - - 6th expedition

# Colonial rivalry

In the decades before Federation, the colonies of Australia had an often-fractious relationship. Like squabbling siblings, they were prone to jealousy and petty feuds. Like adolescents, they agreed on little except a shared resentment of being told what to do by their mother country, Britain.

South Australia, a small colony living in the shadow of its more prosperous neighbours, was anxious to develop trade routes with India and beyond. An international telegraph line would give South Australia competitive advantage, and maintain its position as the communications hub for the continent.

Information is power. The South Australians knew this already. When ships from Britain, carrying the latest (albeit months old) newspapers arrived in Australia, Adelaide was usually the first port of call to have a telegraph link. Newspapers from other colonies had teams of journalists in Adelaide, all waiting anxiously for the next mail boat and fighting to get first use of the telegraph to send their reports.

As early as 1858, Governor MacDonnell had requested finance for an overland telegraph line, but enthusiasm for an international link faded as problems with submarine cables became apparent. A telegraph cable laid across Bass Strait with much fanfare in 1859 turned out to be a costly failure. The first attempt to lay a transatlantic cable, linking Europe and America, also failed, and losses ran in the millions of pounds.

These setbacks were combined with an economic downturn, and for most of the 1860s the colonial governments let the idea rest.

With the end of the American Civil War in 1865, however, the telegraph rollout gained momentum in the United States. A second attempt to lay a cable across the Atlantic was successful. Soon after, Bombay was connected by a cable under the Indian Ocean to Suez, and on to London.

The newly formed British Australian Telegraph Company (BAT) was planning to link Singapore to Batavia (modern Jakarta) and then to run a cable to the Australian mainland. The big question was, which of the colonies would win the connection.

Queensland, only newly separated from New South Wales, had big plans. The superintendent of telegraphy there, W. J. Cracknell, was ambitious and enterprising. He rapidly built telegraph lines north up the coast, reaching Port Denison, just south of Townsville, and then east across the York Peninsula to Normantown. This would be the young colony's gateway to the world, and a huge boost to its economy and status.

By 1870, Cracknell thought the connection was in the bag: but there was one problem. The BAT cable was to come ashore at Port Darwin. This was rival country: control of the Northern Territory had been passed from New South Wales to South Australia in 1863, largely because of Stuart's successful expeditions through the

centre. BAT needed to ask South Australia for permission to build the part of the line that would run through its territory from Darwin to the Queensland border.

The request shocked the South Australian authorities. They saw a link to Queensland as a threat to the future development of their colony, and their key role in telecommunication.

If the Australian colonies at times behaved like squabbling children, there was some benefit. As every parent knows, sibling rivalry has few equals as a spur to action.

After a decade of dithering, the South Australian government found a sudden courage. It offered BAT an alternative plan: South Australia would, at its own expense, erect a telegraph line from Darwin to Port Augusta (already connected by telegraph to Adelaide). It guaranteed that the line would be open for traffic by January 1872. South Australia did not, as is sometimes asserted, refuse BAT permission to build the line to the Queensland border.

However, the South Australian offer was too good to refuse, as it saved BAT the expense of constructing hundreds of miles of land line.

BAT accepted the South Australian proposal. Queensland had been outmanoeuvred, and W. J. Cracknell was bitter about his defeat.

For most of his professional life, Charles Todd rose above petty colonial rivalry. He worked well with his counterparts in Victoria and Western Australia, planning a unified continental telegraph system, using standard technology and codes. However, the race to build the Overland Telegraph showed that Todd, too, had his competitive side.

A bill to authorise and finance the Overland Telegraph—the largest infrastructure project ever undertaken in Australia to that time—was speedily placed before the South Australian Parliament on 8 June 1870. Approved by a large majority, the bill was signed by the Governor just twelve days later.

# 1,800 miles of wire

The logistical problems facing Todd were immense. In order to speed up construction, Todd divided the project into three sections, which would be constructed simultaneously. The southern section ran 500 miles (800 kilometres) north from Port Augusta. Closer to sources of supply and running over land already surveyed, this section was relatively straightforward. The central section was the longest and most difficult, running through 600 miles (960 kilometres) of country of which little was known apart from those parts Stuart had explored. The northern section would be built southward, from Port Darwin. The team constructing the northern section had to sail from Adelaide all the way around the coast to Darwin, carrying aboard everything they would need.

Provisioning and equipping the huge parties who would construct the telegraph was an immense task. The poles holding up the wire were, where possible, to be cut from local timber. Everything else had to be carried in: insulators, batteries, tools—from axes to shovels to wire cutters—food and medical stores, thousands of iron poles for areas where there was no local timber, and 1,800 miles (2,900 kilometres) of galvanised wire, every yard of it imported from England.

Each work party included blacksmiths, carpenters, cooks, storekeepers, linesmen, surveyors and telegraphers. Transporting all these goods and people were dozens of wagons and hundreds of horses and bullocks to draw them. Two camel caravans were also employed in the central section.

Precise and detailed instructions were Todd's hallmark. He tried to leave nothing to chance: the construction teams had clear lines of command and responsibility, and continual inspections to ensure the work was performed well. But Todd was no martinet: he understood the crucial importance of good morale, and worked hard to maintain it.

With the exception of the narrow corridor described in Stuart's journals, much of the central section remained uncharted. Surprisingly, Todd did not employ any Aboriginal guides to assist the work. Even so, his instructions on the most critical task facing those surveying the route—finding water—were detailed and well-informed, drawing on the observations of Stuart and other explorers.

Ironically, it was not a shortage of water but the opposite that caused the greatest difficulties in construction. The party working on the northern section, beginning at Darwin, made good initial progress, erecting some 100 miles (160 kilometres) of wire, when torrential rain began. The country became a quagmire and further work impossible. Faced with a strike by the labourers, section overseer G. R. McMinn panicked, cancelled the contract and sailed for Adelaide.

The submarine cable from Java arrived in Darwin in November. The international link was ready, waiting to be hooked up . . . but there was nothing to connect it to.

The government sent a Department of Works engineer, Robert Patterson, to Darwin with large numbers of labourers and draft animals, and extra supplies. Before Patterson could get work underway, disaster struck again. The wet season broke early and with unusual intensity, with more than 20 inches (500 millimetres) of rain falling in December 1871 alone.

Patterson had to send for more help. To add humiliation to his problems, this could be done only by sending someone in a boat to Norman-ton, and from there sending a telegraph to Adelaide over the rival Queensland network.

The submarine cable from Java arrived in Darwin in November. The international link was ready, waiting to be hooked up . . . but there was nothing to connect it to. South Australia was committed to having the Overland Telegraph working by the new year. After that, the government would be liable to steep penalties. Patterson, a gloomy soul, succumbed to despair. 'Can see nothing but blackness and suffering ahead,' he wrote. 'Fear expedition must collapse.'

But Todd was made of sterner stuff. He arrived to take charge of the northern section in January 1872, with more draught animals and, more importantly, confidence and determination. In April, improved weather allowed work to resume, and for once there was a stroke of luck: the line between Darwin and Java failed, so the threat of a crippling compensation claim no longer hung over the enterprise.

By June the gap in the line was sufficiently reduced that a horse could cover the distance between Daly Waters and Tennant Creek, and for several weeks a pony express ran messages between the two ends of the line, which were closing on each other by the day. It was now possible for a message to travel from England to South Australia in little more than a week.

LANDING THE LINE AT DARWIN, 1871

# LONDON TO ADELAIDE IN EIGHT DAYS!

*ADELAIDE REGISTER*

# Connection

The line was finally joined at Frews Ironstone Ponds on 22 August 1872. Ironically, it was the pessimistic Patterson who made the final join. There is a story, probably untrue, that he received an electric shock in doing so.

For Todd this was a moment of enormous achievement. He made the first official transmission from Central Mount Stuart, the centre of the continent and the memorial to the 'wee Scot', John McDouall Stuart. Near the cairn that Stuart had erected, Todd connected a portable transmitter and sent the first telegrams to Adelaide.

In Adelaide the news was met with jubilation. The bells of the town hall rang, and flags and bunting appeared on official buildings all over the city. Although the international link at Darwin was not restored for another two months, this was a moment of extraordinary significance in Australian history.

Wish to confirm the completion of the telegraph which is an important link in the electric chain of communication connecting the Australian colony with the mother country and the whole civilised and commercial world will, I trust, redound to the credit of South Australia.

CHARLES TODD

# Powering the line

ALICE SPRINGS REPEATER STATION, 1896

Electricity is so much part of our lives now that it is easy to overlook a basic question about the Overland Telegraph. How was it powered? There was no mains electricity, even in the cities: from where did an electric telegraph, stretching 1,800 miles through an arid wilderness, get its electricity?

Every 120–180 miles (200–290 kilometres) along the line was a repeater station. This was a building where an operator would take down signals and then re-transmit them to the next station. These repeaters were necessary because there was a limit to how far a clear telegraph signal could travel along a single uninsulated iron wire.

Each station had its own power supply, a set of glass batteries called Meidinger cells. These were large, cumbersome things, about 10 inches (25 centimetres) tall and 6 inches (15 centimetres) across. They used a chemical reaction between copper sulphate and magnesium sulphate to generate electricity. Each cell produced only about as much electricity as a modern AA battery, so large numbers of them had to be hooked up together to produce the necessary operating power, about 120 volts. The batteries required continuous maintenance if they were to perform adequately.

The line itself was simply No. 8 standard wire gauge galvanised iron wire, not unlike modern fencing wire. There was only one strand: instead of a circuit formed from two separate wires, the wire was earthed at the receiving end. This meant that leakage of current—because of inefficient insulators, or dust build up, or a wet tree branch brushing against the wire—could seriously interfere with transmission.

# Conflict

What is now the Northern Territory has borders that were almost accidental. As the younger colonies of Queensland, South Australia and Western Australia were formed and separated from New South Wales, what became the Northern Territory was the bit left over that no one wanted. Its boundaries gave little heed to geographic regions that might enjoy a natural unity, and none to the existing boundaries of the Aboriginal peoples. About 120 Aboriginal groups had all or most of their land in what is now the Northern Territory, and they recognised a complex set of boundaries and rules for the use of shared resources.

When John Stuart crossed the interior of Australia, he did so in ignorance of these boundaries. His party must have excited fear and wonder among people who had seen neither horses nor white-skinned men before. The Arrernte people of the central desert later spoke of their fear and puzzlement at the Stuart party's tracks.

The Arrernte did not impede Stuart's journey, but farther north, in June 1860, he was attacked by men of the Warramunga people. The encounter, short and bloodless, was immortalised in the name given to the dry river where it occurred: Attack Creek.

What provoked the attack is unknown, but it is likely that Stuart had unwittingly crossed an important tribal boundary.

The people of the inland often were the custodians of vast tracts of land. Though these could encompass thousands of square miles, they could support only a few hundred people. Each group moved in seasonal patterns, usually in small parties, over land that was recognised as theirs by neighbouring groups. Several groups also had shared usage rights to some food and water resources.

Todd, like Stuart, was a humane man. Both gave strict orders that the 'native people' should be treated well, and that force should be used against them only as a last resort. However, ignorance can be as harmful as malice. Todd did not employ a single Aboriginal person in his construction teams. The boundaries of the desert groups were invisible to Todd's surveyors when they planned the route for the Overland Telegraph. The work parties on the line frequently must have made unwelcome, if not intolerable, intrusions.

During the building of the line there were at least two incidents of serious violence. The worst left two men dead: one white, one black. There were probably many more lesser confrontations. In each case, the exact cause of the violence is unknown, but the arrival of more and more white men and their animals must have been a sore provocation. And while Todd strictly charged his men to avoid all contact with Aboriginal women, it would be naïve to expect that this rule was universally obeyed.

Further confrontation was almost inevitable, because the repeater stations had to be located near reliable sources of water. These springs and wells were also of enormous importance to the native people, and the appropriation of these vital resources caused resentment and hostility.

ARRERNTE MEN RUBBING THE EDGE OF A
SPEAR-THROWER ON A SHIELD TO MAKE FIRE,
ALICE SPRINGS, 1896

In February 1874, the repeater station at Barrow Creek was attacked by a group of Kaititja men, probably because it was too close to an important waterhole. A linesman, John Franks, died almost instantly. The station chief, James Stapleton, was hit by four spears and died the following day.

Revenge was swift and terrible. The South Australian authorities despatched a punitive expedition led by Trooper Samuel Gason. It searched the district for two months. Garson made no arrests, and reported that he had killed eleven Aboriginal men. Other sources claim that Gason's party killed indiscriminately, and that the true number of dead was as great as fifty.

Two other telegraph operators at Daly Waters were killed in an attack by Aboriginal people in June 1875. There was no further recorded violence between the indigenous inhabitants and telegraph workers, but these tragic encounters were representative of the social dislocation the wire wrought on the people whose land it crossed.

Charles Todd and those who planned and built the line did not, for the most part, intend any harm, but Todd's pride in his achievement is revealing. He was an engineer, helping build the Empire. He was scarcely conscious of the people who inhabited the country through which the wire ran.

ARRERNTE WOMAN RESTING ON A DIGGING STICK AS SHE DRINKS FROM A WATERHOLE, FINKE RIVER, 1895–97

What kind of creatures could men be who had broad, flat, toeless feet . . .? As for the horse tracks, we could tell that they must have been made by huge four-legged creatures . . . Surely, we thought, both these kinds of creatures must be evil, man-eating monsters.

COMMENTS OF OLD NORTHERN ARRERNTE MEN TO T. G. H. STREHLOW, 1933

No line passing through a similar extent of uninhabited country, where the materials had to be carted over such long distances, no line of equal length and presenting similar natural obstacles, has been constructed in the same short space of time.

CHARLES TODD, 1870

# The magic chain

The Overland Telegraph ran massively over budget. Todd's initial estimate had been 120,000 pounds; the final cost was almost four times that amount. However, while the international telegraph always struggled to pay its way, it was of enormous economic, political and cultural importance.

The telegraph link to the world was, Geoffrey Blainey argues, a revolution more momentous than the beginnings of air transport fifty years later. Australia, still physically remote, could now communicate with the rest of the world at what was astonishing speed. Newspapers carried information on commercial and political events in Europe within forty-eight hours of their happening. The Governor of New South Wales, Sir Hercules Robinson, said that the 'Earth has been girdled with a magic chain'. There was a new sense of economic and political possibility.

The telegraph stimulated a sense of Australian unity, and was an important impetus towards Federation.

In 1874, an explorer from Western Australia was searching for potential grazing land in central Australia when he came upon the wire, curving between its poles, near Charlotte Waters. 'The telegraph line,' he wrote, 'is most substantially put up and well wired and is very creditable.' This chance encounter seems appropriate. The explorer was John Forrest, the man who, decades later, would be responsible for other great engineering projects, and who, by shepherding Western Australia into the federated Commonwealth of Australia, would help build a nation.

# STILL BUILDING

Australia is a nation still under construction. It is possible to pretend otherwise if, like most Australians, you live in one of the large coastal cities. But leave one of the main highways and drive for a few hours, and you will be reminded that Australia is a vast land, and that EFTPOS and sealed roads and mobile phone signals reach only so much of it.

It is a hard land, too, much of it: dry, ancient, worn, the soil lying thin. The country was strange to European settlers, and in attempting to subdue it they made some catastrophic mistakes. They destroyed millions of acres of marg inal land with the cloven hooves of cattle. They discovered and then carelessly wasted the waters of the Great Artesian Basin. Through backbreaking labour they transformed valuable forest into degraded and untenable farmland.

But there was an energy and excitement about colonial Australia. If the nineteenth century was the age of Empire, it was also the age of the engineer. Developing a vast and often inhospitable land like Australia demanded engineering feats of breathtaking scale—a telegraph line crossing a continent; a bridge taller than any structure in Australia and spanning more than 1,600 feet, nearly 500 metres; a pipeline carrying water 300 miles (480 kilometres) uphill across the desert—these were miracles in their day. The men who built them—C. Y. O'Connor, Charles Todd and J. J. C. Bradfield—were public figures, objects of both admiration and fierce criticism.

Each project was about Australia's place in the world. The Overland Telegraph linked Australia to the rest of the Empire, and to the modern industrialising world. The Goldfields Pipeline secured the viability of a valuable mining industry that underpinned the economic future of Western Australia and its role in the new federated nation. The Sydney Harbour Bridge asserted Sydney's status as a world city, and Australia's prominence in the British Empire.

'The achievement of this Bridge is symbolic of the things Australians strive for,' said Jack Lang after the drama of the opening ceremony, 'a bridge of common understanding that would serve the whole of the people of our great continent. That bridge, unlike this, is still building.'

The Great Depression put all thought of major capital works on hold for the best part of a decade, but even before the end of World War II, engineers and their political masters were again thinking big.

The Snowy Mountain Scheme was one result. Legislation for the project, which captured the water of the Snowy and Eucumbene rivers in massive dams and diverted it for electricity generation and irrigation, was passed in 1947. The project was not fully completed until 1974.

The Snowy was an epic engineering feat. Its combined storage capacity was equal to that of every dam and reservoir in Australia built before 1940. A hundred miles (160 kilometres) of tunnels and 80 miles (130 kilometres) of aqueducts carried the water through a mountain range to the inland plains.

In charge of the scheme for almost three decades was William Hudson, another engineer of heroic stature in Australian history.

Quite apart from his professional skills, Hudson got the politics right. Like Bradfield, Hudson was able to sell his massive and costly scheme to the public. Like the vast Hoover Dam on the Colorado River in the United States, the Snowy became an object of national pride, a symbol of nation building.

Big infrastructure projects are highly political. They consume vast amounts of public money. They secure the economic future of some sections of society, and threaten that of others.

They may alter the landscape, and can dislocate entire communities. For indigenous people, a water pipe or a telegraph wire running through their land was, very often, a catastrophe. The social structures and patterns of land use of millennia were destroyed in a generation, with shocking consequences that endure to this day.

For all this, the contribution of engineers in creating the Australian nation was immense.

The skills, values and visions of engineers helped shape our systems of government and administration, the economic basis of the nation, and the physical form of our cities and towns.

Once something is successfully completed, it is accepted, embraced and, very often, taken for granted. All the arguments about whether the scheme will work, whether the cost is justified, whether the people in charge are up to the job, all the doubt, uncertainty and criticism are easily forgotten.

But the battles that are fought in the creation of great projects should not be forgotten. It is important that we learn from our story, and gain a better understanding of what it takes to build something important, and build it well.

# VISUALISING HISTORY

Translating the past to the screen is exciting partly because it's such a risky business.

In books and written documents, as well as in physical remains, we can glimpse history directly. Original letters, documents, images, artworks, buildings, even apparently everyday objects, are all direct routes that allow us to go back in time.

Reading the original words, standing in the same spot, holding the same objects as someone in the past can be powerfully evocative, but these opportunities are rarely available to most of us.

Visual representations of the past these days are far more commonly experienced through the medium of television. There is a variety of approaches, from programs that rely largely on archival imagery, to full-blown historical drama. Along this continuum the accuracy of the history that is presented can vary greatly.

In fictional drama, the past is merely a backdrop to a story, but there is an emerging genre of historical documentaries that employs dramatised reconstructions and 'docudrama' to evoke past characters and events. Ensuring that the past is represented in valid ways and remains within acceptable bounds of historical accuracy, while keeping viewers engrossed, is an important challenge for filmmakers working within this genre.

In *Constructing Australia*, telling these three important stories was influenced by how they could be visualised for maximum effect and the need to relay them to a large audience with as much impact as possible. The impact relies on amplifying the fundamental fascination of the history by presenting it through the characters at the heart of the big-picture events. This is not the same thing as academic investigation, but nevertheless still has to be done according to high standards of research. For the *Constructing Australia* project, that research not only extended to historical events, but also to visual details such as costumes, equipment and locations. The focus was on factual detail and analysis using expert testimony and properly researched narration. In addition, the actors' dialogue was confined only to the actual words that historical records prove the people they are representing spoke. These are drawn from letters, diaries, court reports, newspapers and other contemporary sources. The aim was to present real information and testimony dramatically, rather than making up scenes purely for drama.

As discussed in the introduction, history can be seen as river made up of the 'droplets' of countless individual lives. In *Constructing Australia* the aim was to put flesh on the bones of some of those characters whose lives influenced our own so that we, the audience, can have an encounter with them and in a small way see a little more clearly how the modern world we inhabit evolved.

*Alex West*

The following pages show production stills from the documentary series.

Stills from *The Bridge*, 2006, are copyright Film Australia, photos by Ross Coffey. *The Bridge* was developed with the assistance of the New South Wales Film and Television Office and produced with the assistance of the Australian Broadcasting Corporation. A Film Australia Making History Production in association with Real Pictures.

Stills from *Pipe Dreams*, 2007, are copyright Film Australia, Pipeline Dream Pty Ltd, ScreenWest 2006, photos by David Dare Parker. *Pipe Dreams* was produced with the assistance of ScreenWest and Lottery West, developed and produced in association with the Australian Broadcasting Corporation. A Film Australia Production in association with Prospero Productions.

Stills from *A Wire through the Heart*, 2007, are copyright Film Australia, photos by Simon Stanbury. *A Wire through the Heart* was produced with the assistance of the Australian Broadcasting Corporation and developed with the assistance of the Australian Broadcasting Corporation and the British Broadcasting Corporation. A Film Australia Production in association with Piper Films and the South Australian Film Corporation.

FROM *PIPE DREAMS*

FROM *PIPE DREAMS*

FROM
*A WIRE THROUGH THE HEART*

FROM *THE BRIDGE*

FROM *THE BRIDGE*

FROM *A WIRE THROUGH THE HEART*

FROM *A WIRE THROUGH THE HEART*

# RECOMMENDED READING

## The Bridge

The Sydney Harbour Bridge has been the subject of a small library of books. One of the better recent titles is Peter Lalor's *The Bridge* (2005), but Peter Spearritt's *The Sydney Harbour Bridge* (1982) remains the benchmark.

J. J. C. Bradfield is the subject of a sympathetic biography by Richard Raxworthy, *The Unreasonable Man: The life and works of J. J. C. Bradfield* (1989).

Jack Lang still awaits the biographer who can turn the story of his tumultuous life into a really fine book. Heather Radi and Peter Spearritt's *Jack Lang* (1977), a collection of papers on different aspects of his long career, is interesting but lacks cohesion. Bede Nairn's *The 'Big Fella': Jack Lang and the Australian Labor Party, 1891–1949* (1986) is authoritative, but heavy going for the non-specialist reader. Frank Cain's *Jack Lang and the Great Depression* (2005) is more accessible, and partly fills the gap.

The story of the New Guard is well told in Keith Amos's *The New Guard Movement* (1976). Andrew Moore's *The Secret Army and the Premier* (1989) is a well-written account of the crisis of 1932, and his recent biography *Francis de Groot: Irish fascist, Australian legend* (2005) fleshes out this intriguing character.

The non-specialist reader wanting to understand the economics of the Depression should turn to C. B. Schedvin's *Australia and the Great Depression* (1970) and L. J. Louis and Ian Turner's *The Depression of the 1930s* (1968). An excellent oral history that emphasises the human consequences of the Depression is Wendy Lowenstein's *Weevil in the Flour* (1978).

## e dreams

*Connor: His life and legacy* (2001) by A. G. (Tony)
a detailed and readable full-length biography of
 at engineer. Geoffrey Blainey's *The Golden Mile*
a history of Kalgoorlie that takes a more critical
ctive on O'Connor and the pipeline scheme. For
are of life on the goldfields, see Ian Templeman and
ette McDonald's pictorial work *The Fields* (1988) and
ittington's social history *Gold and Typhoid: Two*
1988).

ward Shann's *An Economic History of Australia* (1930)
describes the economic and political environment of
development. David Ingle Smith's *Water in Austra-*
3) is an excellent introduction to the nature of our
esources and the problems of managing them. To my
as a historian I find Federation a dull affair. For those
g an authoritative yet accessible guide on the subject,
ntenary Companion to Federation* (1999) edited by
rving is recommended.

## A wire through the heart

Tom Standage's *The Victorian Internet* (1998) is an accessible and informative global history of the electric telegraph. Geoffrey Blainey's *The Tyranny of Distance* (1968) is a
classic work of economic and social history that emphasises
the symbolic significance for Australia of the Overland Telegraph. The early chapters of Ann Moyal's excellent history
of telecommunications in this country, *Clear Across Australia*
(1984), cover the development of the telegraph system. Alan
Powell's *Far Country* (1982), a history of the Northern Territory, details the complex political history of the region and
the effects of the wire on the Aboriginal population. For a
detailed account of Todd and the building of the Overland
Telegraph, see Peter Taylor, *An End to Silence* (1980).

*Richard Evans*

# ACKNOWLEDGEMENTS

I have drawn on the work of many fine historians in researching this book. Most of them are listed in the recommended reading, but it would be remiss not to mention also the *Australian Dictionary of Biography*, the indispensable starting point for anyone looking into Australian history. I am indebted to Sybil Nolan and Mark Davis for recommending that I be asked to write this book, and to Elisa Berg and Alex West for taking me on. Thanks to Cathy Smith for her good-humoured editing, to my feisty agent Jenny Darling, and to the indomitable Donica. Writing this book in such a short period would have been impossible without the tolerance and flexibility of my various employers: I am particularly grateful to Richard Broome at La Trobe University. By far the greatest burden, however, was placed on my wife, Heather: for this and for all things, thank you, my love. It is to her, to our two beautiful children, Rose and Zoë, that I dedicate this work. I give thanks to God for the amazing world – so full of stories! – in which we live, and for the opportunity to write about it.

Richard Evans

Melbourne University Publishing and the author are grateful to Film Australia and the documentary filmmakers and their teams for their efforts in helping to produce this book. They are, for *The Bridge*, Simon Nasht and Renee Kennedy; for *Pipe Dreams*, Julia Redwood, Ed Punchard, Rose Grandile and Franco Di Chiera; for *A Wire through the Heart*, Mike Piper, Darcy Yuille, Corey Piper and Mike Dunn; at Film Australia, Anna Kamasz, Jackson Pellow, Paula Bray, Liz Fisher, Lucy Milne, Daryl Karp, Mark Hamlyn, Stuart Menzies, Liz Stevens, Alex McDermott and Martien Coucke; at Postmodern Sydney, John Lau, Christine Wells and Luke Simhauser. We also acknowledge the generous support and efficient service of the State Libraries and their helpful staff, especially Linda Davis at the Battye Library, State Library of Western Australia, Lew Chapman at the State Library of South Australia and Jennifer Broomhead at the State Library of New South Wales.

# CREDITS

**The Bridge:** 2–3: Road Traffic Authority, NSW; 4–6: State Records, NSW; 7: Newspix; 8: State Records, NSW; 9: Harold Cazneaux; 11: Mitchell Library, State Library of New South Wales; 13: Frank Cash; 14–16, State Records, NSW; 17: private collection; 18: Australian War Memorial Negative Number H11576; 19: Australian War Memorial Negative Number E00098; 21–25: State Records, NSW; 27: State Records, NSW; 28–33: Frank Cash; 34: National Gallery of Australia/Harold Cazneaux; 35: Frank Cash; 36: State Records, NSW; 37: National Library of Australia/Harold Cazneaux; 39: private collection; 40: *Bulletin*/ACP; 42–44: National Library of Australia/Harold Cazneaux; 46–47: State Records, NSW; 49: Hood Collection, State Library of New South Wales; 52: Historic Houses Trust; 55–57 (top): Sydney University; 57 (bottom): Historic Houses Trust; 59: Peter Spearritt/Historic Houses Trust; 60–61: Maritime Museum/Annand; 63: State Records, NSW; 65: Film Australia/Ross Coffey; background motifs throughout supplied by Peter Coleman.

**Pipe Dreams:** 66–67: Photolibrarycom/Getty Images; 68: Battye 010738D; 69: National Trust, Western Australia/Forrest Family; 70–71: National Library of Australia; 72: Battye 5182P/000763D; 74–75: Battye 9195P; 76: Battye 1337B/2; 77: Battye 008930D; 78–79: Battye 4144P; 80–81: Battye 4718P/Battye 000632D; 82–83: Battye 9238P/Battye 000877D; 85: Battye 2240B/58; 87: Battye 731B/4; 88–89: Battye 209347/Battye 000808D; 90–91: State Records Office, WA; 93: National Library of Australia; 94–95: Battye 10209P; 96: Battye Library, unknown; 97: Battye—A Splendid Opening, *Western Mail* 7/4/1899; 98–99: Battye Library, unknown; 100–101: Battye 009132D; 102: Battye 13494D; 103 (top): Battye 13456D; 103 (bottom): Battye 013466D; 104–5: Battye 209356; 106: Battye 13505D; 107: Battye 013506; 111: State Records Office, WA; 113: Battye 013466D; 115–17: National Trust, Western Australia/Forrest Family; 119: Film Australia/Julia Redwood.

**A Wire through the Heart:** 120–21: Corbis; 122: photographs courtesy of the State Library of South Australia (Stuart: SLSA B 501; Todd: SLSA B 12209); 123: photograph courtesy of the State Library of South Australia (SLSA: B 22103/37); 126–27: National Library of Australia; 129: Corbis; 130: Film Australia/David Dare Parker; 133: National Library of Australia/Deborah Parry; 135: original map illustrated by Guy Holt; 137: photograph courtesy of the State Library of South Australia (SLSA: B 22663); 138–39: photograph courtesy of the State Library of South Australia (SLSA: B 4639); 140–41: photograph courtesy of the State Library of South Australia (SLSA: B 69996/65); 143: photograph courtesy of the State Library of South Australia (SLSA: B 16); 144–45: photograph courtesy of the State Library of South Australia (SLSA: B8349); 146: Museum Victoria, catalogue no. XP14301, photograph by Baldwin Spencer; 148: Museum Victoria, catalogue no. AP5952, photograph by Baldwin Spencer; 149: Museum Victoria, catalogue no. XP9551, photograph by Baldwin Spencer; 151: iStockphoto.com/Ben Goode.